SHRUBS OF THE
GREAT BASIN

BOOKS IN THE
GREAT BASIN NATURAL HISTORY SERIES

Trees of the Great Basin
Ronald M. Lanner

Birds of the Great Basin
Fred A. Ryser, Jr.

Geology of the Great Basin
Bill Fiero

Fishes of the Great Basin
William & John Sigler

Shrubs of the Great Basin
Hugh N. Mozingo

MAX C. FLEISCHMANN SERIES
IN GREAT BASIN NATURAL HISTORY

SHRUBS OF THE GREAT BASIN

A Natural History

HUGH MOZINGO

*Drawings By
Christine Stetter*

UNIVERSITY OF NEVADA PRESS
RENO :: LAS VEGAS

GREAT BASIN SERIES EDITOR
JOHN F. STETTER

FRONT COVER: Wild Rose. *Rick Stetter*
BACK COVER: Elderberry. *Stephen Trimble*

The paper used in this book meets the minimum requirements of American National Standard for Information Sciences—Permanence of Paper for Printed Library Materials, ANSI Z39. 48—1984.

Library of Congress Cataloging-in-Publication Data

Mozingo, Hugh Nelson.
   Shrubs of the Great Basin.

(Max C. Fleischmann series in Great Basin natural history)
   Bibliography: p.
   Includes Index.
   1. Shrubs—Great Basin.   I. Rasmuss, Christine.
II. Title. III. Series.
QK141.M68   1986   582.1'4'0979   86-7070
ISBN 0-87417-111-3 (alk. paper)
ISBN 0-87417-112-1 (alk. paper : pbk.)

COPYRIGHT © UNIVERSITY OF NEVADA PRESS 1987
ALL RIGHTS RESERVED
COMPOSED AND PRINTED IN THE
UNITED STATES OF AMERICA

*To my wife Katherine
and my mother and father.*

CONTENTS

Acknowledgments	xv
The Great Basin	1
Shrubs in the Great Basin	4
CUPRESSACEAE	
Dwarf Juniper	15
EPHEDRACEAE	
Ephedra	19
BERBERIDACEAE	
Creeping Barberry	27
FAGACEAE	
Bush Chinquapin	32
BETULACEAE	
Mountain Alder	37
CHENOPODIACEAE	
Iodine Bush	43
Four-winged Saltbush	46
Shadscale	52

Torrey Saltbush	60
Saltsage	63
Winterfat	67
Spiny Hopsage	73
Green Molly	77
Greasewood	80
Desert Blite	87

POLYGONACEAE

Kearney's Buckwheat	92
Great Basin Buckwheat	96
Rock Buckwheat	100

TAMARICACEAE

Tamarisk	106

SALICACEAE

Coyote Willow	113

BRASSICACEAE

Bush Peppergrass	119

ERICACEAE

Greenleaf Manzanita	123
Western Blueberry	129

GROSSULARIACEAE

Western Golden Currant	134
Wax Currant	137
Plateau Gooseberry	141

ROSACEAE

Western Serviceberry	145
Littleleaf Mountain-Mahogany	149
Fern Bush	154

Blackbrush	157
Cliffrose	159
Ocean Spray	162
Dwarf Ninebark	166
Desert Peach	168
Bitter Cherry	172
Bitterbrush	175
Wild Rose	183

FABACEAE

Smokebush	187

ELAEAGNACEAE

Silver Buffaloberry	195

CORNACEAE

American Dogwood	199

CELASTRACEAE

Spiny Greasebush	204

RHAMNACEAE

Tobacco Brush	208
Sierra Coffeeberry	214

ACERACEAE

Dwarf Maple	218

ANACARDIACEAE

Squawbush	224

SOLANACEAE

Shockley's Desert Thorn	229

POLEMONIACEAE

Prickly Phlox	234

LAMIACEAE

Purple Sage	238
CAPRIFOLIACEAE	
Twinberry	243
Elderberry	247
Snowberry	251
ASTERACEAE	
Dwarf Sagebrush	256
Silver Sagebrush	263
Bud Sagebrush	265
Big Sagebrush	270
Littleleaf Brickellbush	283
Rubber Rabbitbrush	287
Parry's Rabbitbrush	293
Green Rabbitbrush	296
Snakeweed	301
White Burrobush	304
Gray Horsebrush	308
Littleleaf Horsebrush	310
Shortspine Horsebrush	315
Cotton Horsebrush	317
Indian Names for Great Basin Shrubs	319
Bibliography	321
Index	331

ACKNOWLEDGMENTS

No one can write an eclectic work such as this without recognizing that, to a great extent, only a small portion of it represents what can be said to be, by any stretch of the imagination, an original contribution by the author. Consequently, this volume represents the distilled experiences of a plethora of people—field-trip companions, my colleagues here in the Great Basin, other researchers—only a small minority of whom I got to know personally, and most importantly, the many students I've led on field trips for twenty-six years who have taught me more than they may realize.

Having said that, however, I would be especially remiss if I did not specifically acknowledge those individuals to whom I owe a particularly large debt, both for the production of this work and for whatever understanding I may have of the natural environment of the Great Basin.

Foremost among these would be Fred Ryser and the late Ira LaRivers, who first introduced this naive easterner to the wonders of the intermountain west. Fred Ryser has been my usual and indefatigable companion for two and one-half decades of peregrinations over the Great Basin (and indeed as far as Australia), and it was largely due to his encouragement that this work was undertaken.

During the seventies, many of our field trips also included Peter Herlan, late director of biology at the Nevada State Museum, whose broad knowledge of both plants and animals added significantly to my enjoyment of these excursions.

Notable among my graduate students who assisted in botanical forays were Ralph Bertrand, Phyllis Henderson, Richard Holbo, Emily McPherson, Ann Pinzl, Irwin Ting, and Richard Trelease. One graduate student, in particular, deserves special mention—John Kartesz. He has been a thoughtful field companion and intellectual adversary during the last six years. His knowledge of the particulars of botanical nomenclature is second to none, and his advice has been invaluable in making certain that my scientific names are not too far out of date, though it should be assumed that any errors the reader may discover exist because I unwisely chose not to follow his counsel on all such matters.

Over the years hundreds of undergraduates contributed to my education, not only through their own intimate knowledge of our flora, but perhaps more through an inquisitiveness that helped to stimulate my own curiosity. To name them all would be impossible since there were so many and my memory is all too fallible. I am very indebted to Donald Baepler, former chancellor of the University of Nevada system, and Robert Laxalt, former director of the University Press, for their expression of confidence in selecting me to write this book. I hope that their support will not prove to have been misplaced.

Two Agricultural Research Service scientists, James Young and Ray Evans, have been especially helpful to me in long discussions that improved my insight into the ecology of many of our range shrubs. Discussions with other botanists have been of enormous help, particularly those with Joe Robertson, Richard Eckert, Kenneth Genz, Ronald Lanner, Paul Tueller, Wayne Burkhardt, Ed Kleiner, Ham and Pat Vreeland, and Fritz Went. Life-science librarians Ann Amaral, Betty Hulse, and Dorothy Good have been of invaluable help in my searches for the more obscure references.

I am also indebted to Kay Fowler, associate professor of anthropology at UNR, for the authorship of the appendix listing Indian names for some of our Nevada shrubs.

If this book reads reasonably well, with a minimum of stylistic and outright grammatical errors, most of the credit should go to Holly Carver, my exemplary copy editor. The Fleischmann Natural History Series Editor and present director of the University of Nevada Press, John F. "Rick" Stetter, is largely responsible for the consistent support without which this effort would never have been completed, given my proclivity for rationalization and procrastination. His wife, Christine Rasmuss Stetter, produced the excellent detailed, yet aesthetically pleasing, line drawings, which should satisfy both the artist and the most discriminating botanist. Aside from my own few and rather feeble efforts, the color photographs, which, I believe, add considerable academic as well as artistic value to this work are those taken by Rick Stetter, Stephen Trimble, and John Running. Nicholas Cady and Cameron Sutherland of the University Press contributed significant editorial assistance along the way.

Finally, I must express to my very patient wife, Katherine, my deep appreciation for enduring the many absences that my professional efforts have occasioned over the years. Without her forbearance and loving support this work would not have been possible.

No portion of this whole district, however desert in repute and in fact, is destitute of some amount of vegetation even in the driest seasons, excepting only the alkali flats, which are usually of quite limited extent. Even these have frequently a scattered growth of *Sarcobatus* or *Halostachys* surmounting isolated hillocks of drifted sand, compacted by their roots and buried branches. The vegetation covering alike the valley plains, the graded incline of the mesas, the rounded foothills and mountain slopes, possesses a monotonous sameness of aspect and is characterized mainly by the absence of trees, by the want of a grassy greensward, the wide distribution of a few low shrubs or half-shrubby plants to the apparent exclusion of nearly all other growth, and by the universally prevalent gray or dull olive color of the herbage.

—Sereno Watson, Botany, U.S. Geological Exploration of the Fortieth Parallel, 1871

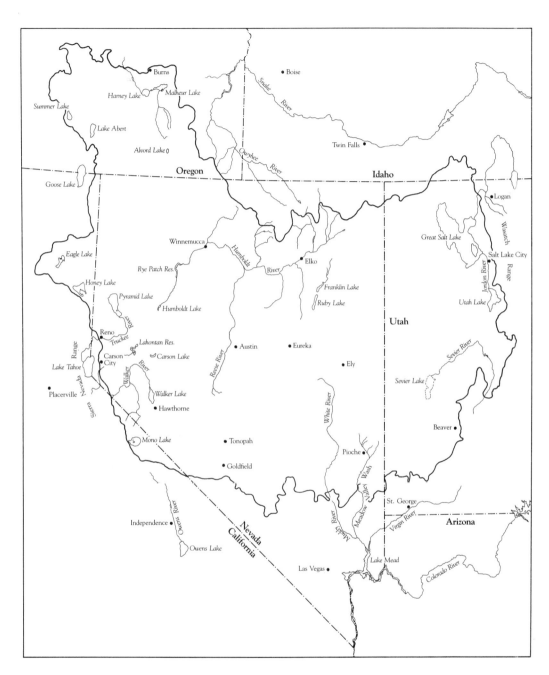

The Great Basin.

THE GREAT BASIN

JOHN C. FRÉMONT, the explorer, coined the name Great Basin for that vast series of north-south, parallel mountain ranges and basins extending from the Wasatch Range in Utah to the Sierra Nevada of Nevada and California. Northward the boundary lies along the Snake River drainage area, including much of southeastern Oregon. On the west the boundary follows the crest of the Sierras and, at the southern extremity of the White Mountains, swings eastward to exclude the southern part of Nevada. The accompanying maps indicate primarily, though not entirely, a somewhat smaller area called the hydrographic Great Basin, an area into which all streams flow and from which no river exits to the sea. Obviously, Frémont should have made his appellation plural, since there are at least seventy-five separate basins, rather than a single basin. And, in reality, the Great Basin has a domed rather than a concave profile—the highest valley floors at the center range from 5,500 to 6,300 feet in elevation, while those to the east and west may be as low as 3,800 feet. Many of the ranges are 6,000 to 7,000 feet high, and a few exceed 10,000 feet, with the highest being the White Mountains at an elevation of over 14,000 feet.

Dwight Billings divided the Great Basin into a number of vegetation zones, which we should note in passing, albeit not in any detail. Except for the extreme saline areas, desert playas, and some lava fields which have no higher plants, the lowest zone is the shadscale zone, named after its most abundant

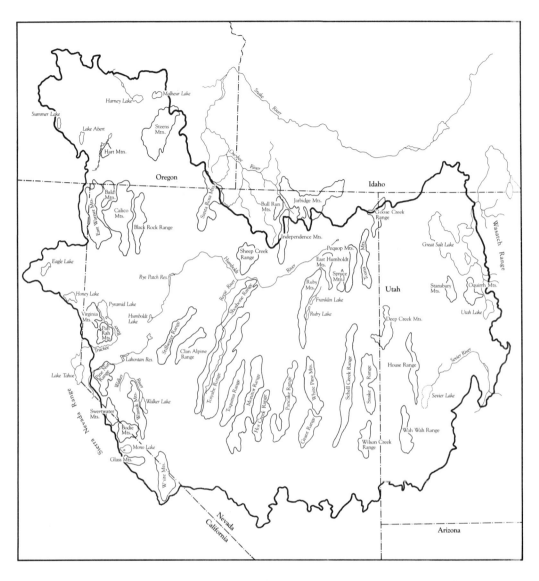

The Great Basin, showing principal mountain ranges, rivers, and lakes.

shrub. Common companions of the shadscale are green rabbitbrush, bud sagebrush, four-winged saltbush, spiny hopsage, Bailey's greasewood, and littleleaf horsebrush. The more saline areas within the shadscale zone include big greasewood as the dominant shrub, along with desert blite and green molly. Sand dunes are likely spots for smokebush and cotton horsebrush.

The sagebrush zone is slightly higher than the shadscale in elevation, with a resultant greater annual precipitation and a lower rate of evaporation. This zone is, in reality, a sagebrush-grass zone, although in many areas grass has largely disappeared because of overgrazing. Along with big sagebrush are found green and rubber rabbitbrushes, blackbrush in the southern part of the Great Basin, green ephedra, bitterbrush, spiny hopsage, and snowberry. Some sagebrush communities extend upward even to mountaintops and elevations over 10,000 feet high.

Above the sagebrush zone, the pinyon-juniper zone is dominated by these two trees. The shrubs within this zone commonly include big sagebrush, snakeweed, bitterbrush, plateau gooseberry, western serviceberry, snowberry, and green and rubber rabbitbrushes.

Still higher is a series of coniferous forest zones: the yellow pine–white fir zone, the lodgepole pine–mountain hemlock zone, and so on. Some shrubs occur in all these zones, including such species as greenleaf manzanita, sierra coffeeberry, various currants and gooseberries, and bush chinquapin.

For the reader interested in a more detailed summary of our vegetation zones, the best place to start is volume 1 of *Intermountain Flora*, by Arthur Cronquist, Arthur Holmgren, Noel Holmgren, and James Reveal. In the discussions to follow, some of the ecological preferences of our shrubs will be considered, along with some of the reasons why certain species are found where they are.

SHRUBS IN THE GREAT BASIN

THE FIRST SHRUBS probably appeared on the earth about 400 million years ago, during the era that geologists call the Paleozoic. Calling these first land plants shrubs is certainly a very liberal use of that word, for they only remotely resembled most of those found today. They were, in fact, only indirect ancestors, since none had flowers and all were structurally quite primitive. But they were the ancestors of the ferns and fern allies which were eventually to evolve into the higher plants of today. These first land plants were poorly adapted to their new environment, for they could grow only in marshy places. Nevertheless, they were the first production models of a successful land plant form that, with some modifications through the process of evolution, has proven to be the best solution the plant kingdom has yet designed for a higher plant which can flourish in the widest range of habitats from the Arctic to the Antarctic. Shrubs can be found everywhere from the rainforest to the desert, from the coldest tundras to the most benign equatorial shores.

Basically, there are three forms of higher plants: trees, shrubs, and the so-called herbaceous types. Simply put, the latter are those forms which lack the ability to produce any woody tissue. Everyone knows what a tree is, and even those botanists who delight in being abstruse would agree with the notion that trees usually have a single main axis, or a trunk, and that shrubs usually possess a profusion of stems, many of which spring directly from the ground at the base of the plant.

However, nature pays no attention to categories invented by humans, and any perceptive observer could point out that the dividing line between trees and shrubs is not always very distinct. Some true trees, such as the Siberian elm, will commonly send up numerous shoots from old stumps and thus come to resemble shrubs, while some undoubted shrubs, such as the mountain alder, look much more like trees as they grow older.

Trees and shrubs, as we pointed out, are woody and have in common a layer of tissue beneath the bark, known as the cambium, which is capable of increasing the stem's diameter by the production of additional wood. There are only a few exceptions to this rule (such as the palms), and none of them occurs naturally in the Great Basin. The lack of a significant cambium and the consequent lack of woody stems characterize the herbaceous plants, which we now believe are later evolutionary creations from shrub ancestors.

The earliest flowering plants were woody, and the tendency to evolve herbaceous types occurred independently and at different times in various families to produce such forms as corn, dandelions, and tomatoes. After this, it was only a short evolutionary step from the long-lived or perennial herbaceous forms to annual types which germinate, grow to maturity, flower, and fruit all within a few weeks or months.

Shrubs occupy a greater variety of land habitats than virtually any other plant form. Why this is so is difficult to explain, at least partially because we may not completely understand this adaptation. The explanation, in fact, would vary depending on which plant community we happened to be examining. The Great Basin has a large number of very successful herbaceous forms, annual as well as perennial, but the really conspicuous and characteristic plants on our enormous vistas of both desert and steppe are the shrubs—so much so that we talk about the shadscale desert, greasewood association, or big sagebrush community. Shrubs are our constant companions here; basically, this is because the major factor limiting plant growth is the relative lack of water. Trees take in more water and evaporate more, and except for our mountain ranges, towns, and waterways there is simply not enough water to allow them to survive, let alone grow. From an airplane, in fact, many of our cities appear to be the only forests at low elevations. Some forestry schools have recognized the reality of this by introducing urban forestry programs.

There are many features that our shrubs share which, when understood, help explain how they are able to survive the rigors of heat, cold, and dryness so much better than trees. What may not be apparent is that virtually every adaptation exacts a price (in nature, as elsewhere, you can't get something for nothing). Most frequently, this price involves a slower growth rate and/or a lower rate of food manufacture. Most of our shrubs, for example, have relatively small leaves. This helps reduce water loss, since less surface is exposed to the dry air. But the price paid is a slower growth rate and a smaller final size. Additionally, the leaves frequently are leathery or somewhat thick and succulent. Some shrubs have a thick, varnishlike layer on their leaves, the cuticle, which cuts down on water loss. Succulent types, especially those with leaves in the form of a thick cylinder, have improved their chances by reducing the surface area–to–volume ratio of the leaves. Generally, a reduced proportional surface area means a lowered transpiration rate, but the price paid is again a slower growth rate. In general, succulent plants which approach a cylinder or sphere in shape have very slow growth rates.

Water loss through the leaves is characteristic of all land plants. No land plant is impervious to at least some evaporative loss, basically because there is no way (insofar as plants have discovered) to allow the carbon dioxide necessary for photosynthesis to diffuse into the leaf without losing some water vapor to the air at the same time. Although plant physiologists still debate its importance, at least some cooling of a sunlit leaf is the result of transpiration. Transpiration is thus a necessary evil if plants are to survive and grow.

In addition to the cuticle, many Great Basin plants have a waxy layer on their leaves. Frequently this waxy layer is whitish; this has the added advantage of reflecting a great deal of light, which would otherwise add to the heat load of the leaf and increase transpiration. However, many of our grayish or whitish desert shrubs appear so because of a layer of fine hairs on the leaf surface. These hairs help reduce water loss by creating a boundary layer of air with a higher humidity content immediately above the leaf surface. This presents an obstacle to water loss and, as might be guessed, is especially important when the air is moving, since even a slight breeze will drastically increase evaporation.

We need to say a little more about the nature of water loss in plants for the reader to fully understand an additional adaptation seen in some of our native shrubs. Transpiration in plants occurs primarily through the leaves, for several reasons. First, the leaves are thin, laminar structures, even if they are small, and thus present relatively more surface area to the air. Second, most water loss (generally above 90 percent) takes place through specialized pores called stomates, which in the vast majority of plants are open during the day and closed at night. Stomates must be open during the daytime if the carbon dioxide essential for photosynthesis is to enter the leaf. Naturally, plants cannot carry on photosynthesis at night, so it makes sense that they should have evolved a mechanism that would allow the stomates to close at night and thus conserve water.

We must beware of a danger in this kind of interpretation, however, for evolution is not a conscious process, and to assume that something exists because it makes sense not only short-circuits our attempt to understand nature but may, on occasion, lead us to a totally erroneous conclusion. To put it simply, desert shrubs do not produce thorns in order to avoid being grazed upon, though this ultimately may be the result—they produce thorns because they have the right combination of genes to do so. Those shrubs in the past which did not deter grazing, by means of thorns or in some other way, in many instances did disappear because of grazing. Why didn't they produce thorns? Their luck just ran out! Put another way, no one proposes that holes develop in roads in order to break the axles of cars, though that may be the ultimate result!

However, some plants—notably cacti and many other succulents—have evolved a striking solution to the problem of getting enough carbon dioxide while reducing water loss to a bare minimum. Very simply, they open their stomates only at night, when evaporation is naturally less, and somehow trap the carbon dioxide which diffuses into the leaf at the same time. The carbon dioxide is trapped during the night because it combines with an organic acid which in effect stores the gas until daylight. Then the process is reversed, and the carbon dioxide thus freed can be used in photosynthesis—it will not escape, since the stomates are closed. Consequently, plants with this physiological ability (known as crassulacean acid metabolism) are ideally adapted to dry habitats, since they significantly reduce transpiration by opening the stomates only at night and at the same time effectively bank carbon dioxide for later use.

Another way of coping is shown by those shrubs which have the stomates sunk in pits beneath the surface of the leaf. This arrangement helps in the same way as hairs to increase the thickness of the stationary air layer through which water vapor must diffuse, and consequently it helps slow down transpiration. A somewhat similar adaptation is shown by the tobacco brush, which curls its leaves toward the underside, where the stomates are located, when dry periods occur.

Other plants have evolved a different method of contending with certain environmental stresses. Desert annuals, for example, germinate, grow, flower, and set seed during the spring and early summer, completing the life cycle before the hottest and driest part of the year. Such shrubs as the littleleaf horsebrush and bud sagebrush, which drop their leaves before the driest season, have a similar life-style. As one might expect, these shrubs are also among the first to develop leaves in the spring. The smokebush is particularly opportunistic in that it produces a new crop of leaves after heavy rainfalls, regardless of when they occur during the spring or summer.

Yet another adaptation exists in the spiny greasebush and the ephedras, which have very small or virtually nonexistent leaves. Their green stems contain chlorophyll and carry on the essential photosynthesis. Obviously, with such a reduced surface area for photosynthesis, the ultimate consequence is a relatively reduced growth rate.

Within the broad valleys separating our mountain ranges in the Great Basin, water, not infrequently, may persist well into summer on the saline playas. But relatively few plants, despite millions of years of evolution, have evolved a solution to the problem of too much salt. In fact, the centers of most of the larger playas are still devoid of any higher plants, though this may not be due entirely to the higher saline content. However, some of our native vegetation, such as saltgrass, greasewood, and iodine bush, not only tolerates high salt concentrations in the soil but actually seems to require saline soil for good growth. Occasional forms, such as the tamarisk, have developed specialized salt glands on the leaves which enable them to get rid of excess salt by the evaporation of water. This deposits the salt crystals on the leaf surface, where less harm will result.

At one time it was thought that the water in saline soils was unavailable to plants, because the salt content was higher than that in the root cells of the plants. It was assumed that this made a net movement of water into the plant physically impossible. A simple explanation, perhaps, but one which now appears to be no more than partial. Nature is almost never as simple as it appears to be (or as we would like it to be)!

Much of the problem with saline soils now appears, on the basis of recent research, to be the result of the toxicity of various common elements and the way in which they interfere with normal physiological processes in plants. Even those elements essential to plant growth are toxic at high concentrations. Some saline area plants or, as they are usually called, halophytes have developed specialized cellular mechanisms which apparently have the ability to keep certain toxic elements from accumulating to high levels within the plant. Whether this is true of most halophytes, however, is not yet known.

Despite the vicissitudes of climate and soil, the natural aspect of the Great Basin appears to the casual visitor to be almost timeless and unchanging. This is certainly not the case, however, at least on the geological scale of measurement (or, for that matter, even on a more recent historical scale). As recently as twelve thousand years ago the lakes formed as a consequence of the last ice age had reached their maximum extent. One of these, Lake Lahontan, was larger than the present Lake Superior of the Great Lakes. Many of the lower-elevation valleys were covered with water and the climate was cooler than it is today, with the result that the areas now occupied by sagebrush and grass were instead conifer forests.

Within a thousand years following the highest lake levels, there began a gradual warming trend which climatologists think peaked about six or seven thousand years ago. From that point to about 3,000 to 2,000 B.C., a decline to our present cooler climate occurred. It seems likely that many of our shrubs now restricted to southern Nevada then ranged well into the Great Basin. We can assume that the shadscale desert and sagebrush-grassland communities began to occupy their present locations about four thousand years ago.

About 50 million years ago, the Great Basin, at least in the western portion, was only a little above sea level and sloped to the west. The warm climate and greater rainfall at that time supported a dense conifer-hardwood forest much like that of northern coastal California today. Some of the shrubs present were alder, serviceberry, gooseberry, and Oregon grape.

Next, there began a lengthy period of mountain building, and the climate gradually became drier. By 13 million years ago the Sierra Nevada was still no more than 300 feet high. Apparently, it was heavily forested and dominated by *Sequoia*. Possibly about this time sagebrush first appeared in the Basin, although the species were most certainly not the same ones occurring today—on an evolutionary scale, a lot can happen and probably will happen in 13 million years! As the climate and topography change, populations migrate and they also inevitably evolve.

Then, between 13 and 1 million years ago, mountain building accelerated and the Sierra Nevada achieved its present elevation. This cut off most of the wet storms from the west, and the flora of the Great Basin had to adapt to very significantly colder and drier conditions. Fossil evidence from the watercourses of that time indicates that chokecherries, poplars, and willows grew along them, much as they do today.

About 1 million years ago, the great Ice Age began. Four major periods of glaciation, with short, warmer interglacial periods, occurred. Although enormous ice sheets were present elsewhere in North America, only a few small glaciers developed on the mountain ranges of the Great Basin, and the flora, consequently, was much more influenced by the numerous lakes which developed at the peak of each glacial epoch. Probably the climate was colder and wetter as a consequence.

By analogy, on the geological scale, one could say that our present Great Basin flora represents but one frame of a constantly changing motion picture. The future aspect of the Basin is certain to be different from that of today, although no one can say in what way. Also certain is the fact that our influence will be as important as were those major topographic and climatic changes of the past.

While we may have difficulty visualizing the changes which take place over hundreds or thousands of years, the very rapid changes due to the natural agency of fire are all too evident at the time they occur. Because uncontrolled fire is so destructive to human affairs, we tend to think it must be just as disastrous and unwanted in nature. Actually, fire before the coming of humans was a common occurrence in most plant communities. Only a few never experience fire, such as certain rainforest types and some swamps.

Some desert scrub communities in which the individual plants are widely spaced, as is the case in the drier portions of the shadscale community, rarely experience fire.

At the other extreme are those communities which not only experience fire annually or oftener but in which the structure of the community and the persistence of individual species are vitally dependent on fire. In the latter case, by attempting to prohibit fire, we are creating an unnatural situation which is as destructive to the environment in the long run as a bulldozer is. Lodgepole pine, knobcone pine, and certain chaparral associations are to a great extent regulated by fire and dependent on it. Many chaparral communities consist of equal-age individuals which date back to the last fire.

A variety of adaptations allow many shrubs to survive even relatively severe fires. Many are stimulated to send up sprouts from their roots within a few weeks after a fire. The growth of these new shoots also seems to be accelerated by the effects of the fire. Some shrubs have fire-resistant bark which can protect the sensitive cambium to some degree. Others have fruit pods which open to release seeds only after a fire, so that a new generation is assured even if the old one is destroyed.

However, not all shrub communities are so resistant to fire. Sagebrush does not readily root-sprout, and bitterbrush is typically destroyed fairly effectively by our range fires. Eventually, both these shrubs may repopulate an area by means of seeds buried in the soil, but this is by no means always certain.

All of the foregoing is but a very brief introduction to the remarkable series of adaptations evolved by our desert shrubs to cope with the vicissitudes of the Great Basin. These adaptations, which involve both function and form, will be explored in somewhat greater depth when we discuss individual species. Unfortunately, a book three times the length of the present one would be required to do justice to the natural history and diversity of all our desert shrubs. So, I have had to compromise, but I have tried to tell something about the natural history of all of our common shrubs and some of the rarer ones. Readers may be aware of other interesting facts which, perhaps, should have been included, or they may well disagree with some of the studies I have summarized. I hope, at any rate, that the stories to follow will be long enough to be both enlightening and interesting, yet not so long that readers suffer from the ennui that all too often accompanies a scholarly treatise—which, it should be emphasized, the present volume is definitely not intended to be. If readers are stimulated to look more closely at the tremendous diversity within our Great Basin shrubs and want to read more about them, then the purpose of this book will have been served.

Finally, I hope that it will be apparent that Sereno Watson, though trained as a botanist, suffered from a myopia common among easterners when they first glimpse our mountains and deserts. Perhaps, to an extent, we all have this problem with a provincial attitude shaping the things that we see, for I can recall some caustic remarks made by a westerner about the monotony of the wooded hills and valleys of West Virginia! Major differences tend, at first, to be more apparent than the diversity within each area. The astute observer, however, soon recognizes that all of the floristic areas on the earth have their own unique and fascinating qualities making them worthy of study by expert or amateur.

CUPRESSACEAE
CYPRESS FAMILY

Dwarf Juniper

Dwarf Juniper
Juniperus communis

IT IS DIFFICULT to visualize the mountains of the Great Basin without also thinking of the apparently limitless expanse of pinyon-juniper woodlands which cover much of their topography. However, far above the Utah juniper so abundantly present is another species of juniper, the common or dwarf juniper—called common because its distribution is truly worldwide in the cooler climates of the northern hemisphere. Botanists characterize the dwarf juniper as circumpolar, since it is found from Alaska across Canada to Newfoundland, in Greenland, and across Europe, northern Asia, and Japan. In the United States it ranges south in the eastern states as far as North Carolina, in the Rockies it extends to New Mexico, and in California it is found south to the central Sierra Nevada. Within the Great Basin it occurs on the higher ranges from 7,500 feet to their summits.

Typically ranging from 30 to 50 centimeters high, the dwarf juniper forms low mats not uncommonly several meters in diameter. Its twigs are somewhat yellowish, have three angles, and possess needlelike leaves, 1 to about 1.5 centimeters long, in whorls of three or five at each node. One curious fact about many junipers, particularly the tree forms, such as the Utah juniper, is that the juvenile leaves and sometimes those on an occasional older twig are needlelike, while those on the mature twigs are small, scalelike structures. Because they are smaller and so closely appressed to the twigs, these scalelike leaves are a superior adaptation to the frequently very dry conditions in pinyon-juniper communities. The scale leaves found in juniper and a number of other cone-bearing trees are undoubtedly evolved from the more common needlelike leaves. It is not at all unusual for juvenile species of many plants to develop leaves typical of those of their ancestors, before the characteristic mature form appears. This juvenile trait, under certain conditions, may be very persistent, as in the case with the common philodendron grown as a house plant. At any rate, we can say that the dwarf juniper remains in a permanently juvenile condition, or, at least, it has never been seen to produce the scale leaves of other junipers.

Like the ephedra or Mormon tea, the juniper belongs to that group of seed-producing but nonflowering plants known as the gymnosperms, a name which means naked seeds, since the seeds during their development are not enclosed in the saclike structure (the so-called ovary, a misnomer if there ever was one) of the flowering plants. The dwarf juniper usually has the male and female structures on different plants, but sometimes they can be found on the same individual. The stamens are borne in small cones about 5 millimeters long, while the ovulate (seed-bearing) cones are spherical structures up to 9 millimeters in diameter. To the uninitiated, they appear to be berries. However, in junipers, the individual scales in the female cone have become fleshy and fused together, and although it may look like a berry it is not at all analogous to the berries formed by flowering plants. Eventually these juniper "berries" become dark blue, with a waxy bloom on them like that of blue grapes. This fleshy juniper cone will contain one to three seeds.

The extract from the berrylike cones of the dwarf juniper was once considered an important spice for preserved meats in the Old World, and it is still used to give gin its characteristic flavor. The English word gin comes from the Netherlands, where the term geneva was applied to this variety of distilled beverage. Geneva in turn stems from *jenever*, which is the Dutch term for juniper. Great Basin Indians used an extract from the twigs as a blood tonic, and in the Old World extracts from juniper were used for a number of ailments involving the kidneys, stomach, and rheumatism. However, Charles F. Millspaugh, in his classical work *American Medicinal Plants*, warns that the extract can be fatal in relatively small quantities.

As we will see in the case of certain other shrubs, a wide distribution, such as the dwarf juniper has, almost inevitably means a lot of variation. Horticulturists and botanists have described many forms of juniper, some of these known only in cultivation. The variety which occurs in the Great Basin is known as variety *depressa*. A different variety, *montana*, occurs in the Sierras, the Cascades, and the Coast ranges. The Great Basin variety is the only one throughout the Rockies, Canada, and the eastern United States.

Juniperus is the old Latin name for juniper, and the species name *commu-*

nis obviously means common. The juniper belongs to the cypress family or Cupressaceae. This family includes our native incense cedar as well as arborvitae and the true cypresses. Altogether the family contains about 16 genera and 140 species. There are about 70 species of juniper, which makes it the biggest genus in the family. All of them are northern hemisphere plants, although some species get as far south as Mexico and the West Indies.

EPHEDRACEAE
JOINTFIR FAMILY

Ephedra
Ephedra spp.

IN GENERAL APPEARANCE, the ephedras are broom- or rushlike shrubs up to 1 meter tall, with noticeably jointed and fluted stems. The leaves, which have evolved almost to the vanishing point, are represented by small, scalelike structures opposite one another at each node. Some forms have three at each node. Photosynthesis is carried on by the green stems. Because of its unique appearance, the ephedra is sometimes referred to as jointfir or shrubby horsetail, though it is not at all related to the true horsetails.

Various ephedras have been called Mormon, Brigham, or settler's tea. Making a medicinal tea from this plant was a practice undoubtedly copied from the Indians—a recent innovation in health food stores has been the packaging of dried stems of ephedra under the name squaw tea! Several Asiatic species of ephedra have been used for over five thousand years in China for the treatment of asthma and hay fever. Ma-huang, as the ephedra is known there, is the source of the drug ephedrine. Our species, however, apparently have little or none of the drug present in their tissues.

E. viridis has bright green stems, as the Latin name implies, while E. *nevadensis* has characteristically gray-green stems. The two scalelike leaves at each node are deciduous in the Nevada ephedra and persistent in the green ephedra. Otherwise the two species are very similar. There are, however, some physiological differences: the Nevada ephedra is a plant of drier desert habitats, while the green ephedra prefers moister locales and is common in the sagebrush and pinyon-juniper zones.

Ephedras do not really have flowers, for they belong to that major group of nonflowering but seed-bearing plants called gymnosperms, which includes, among other things, pines, firs, junipers, and tropical cycads. Ephedras have the stamens borne in small cones at the nodes of younger branches. Each stamen has one to eight pollen sacs. Surrounding the stamens is a calyxlike structure, which superficially resembles a small, greenish flower. Pollen is distributed by the wind. There are no pistils; instead, the one or two ovules are exposed at the tips of small, scaly cones. These ovules, after pollination and fertilization, develop directly into seeds. Superficially, the ovule resembles a pistil as a consequence of the inner of the two surrounding tissue layers extending an elongated, stylelike pollen chamber through an opening in the outer layer at the upper tip of the ovule. Male and female cones usually develop on separate plants, although, on occasion, they will form on the same plant. When the sexes are produced on separate plants, botanists call such individuals dioecious. If, however, stamens and pistils are formed in separate flowers but on the same plant, they are then said to be monoecious. A good example of this latter type would be corn, which has only female flowers in the ear and only male flowers in the tassel at the top of the stalk.

A very intriguing study on the green ephedra and four other species of plants which have male and female structures on separate individuals was carried out by D. C. Freeman, L. G. Klikoff, and K. T. Harper. In the case of green ephedra, male plants were "found in greater numbers on steep slopes," while females were "more common on better-watered sites at the base of the slopes." There are at least two theories which might explain this unequal distribution of male and female plants. A Russian study on plants with dioecious flowers concluded that male plants were less affected by water stress than were female plants in the six species which were examined. So perhaps seeds potentially capable of producing male or female plants are formed in equal numbers, but a differential survival of the two types occurs. Here, one has to assume that sex is genetically fixed, as in humans, and that the higher ratio of females on the moister sites is the result of some competitive advantage which they have. An alternative explanation would be that sex is not genetically fixed but is to some extent under environmental control. For example, some years ago, a study on the eastern jack-in-the-pulpit showed that young plants produced only male flowers, while mature plants with more food reserves in their rhizomes produced female flowers. However, if part of the rhizome was removed the plants reverted to producing only male flowers. Another study by a Russian investigator on marijuana showed that a number of physiological differences existed between male and

Ephedra female flowers

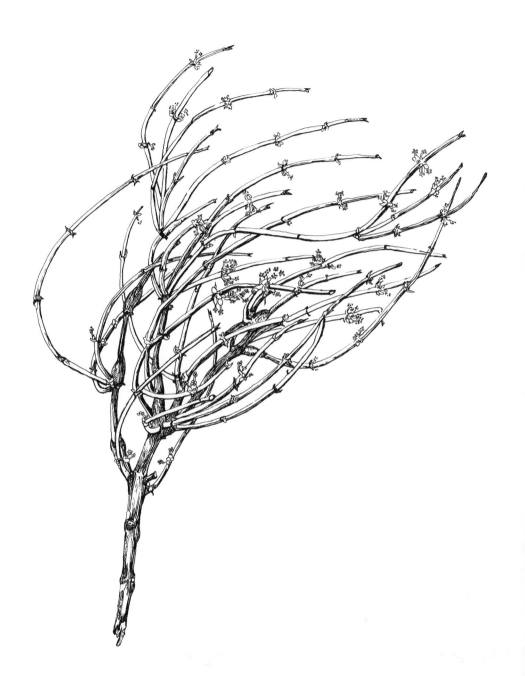

Ephedra male flowers

female plants and that, surprisingly, by treating marijuana seeds before germination with certain organic compounds it was possible to alter the ratio of females to males so that many more female plants were produced.

We really don't know which of these two theories is the correct explanation for the situation in ephedras, but the first would have a higher cost-benefit ratio in controlling sexual expression since it would involve the death of many seedlings, either male or female, depending on the site. In any event, the differential distribution of the sexes in this fashion, according to Freeman and his coworkers, may insure a better distribution of pollen since the males are on the windier sites. Additionally, it takes more energy, nutrients, and moisture for a longer period to produce seeds than it does to produce pollen, so the moister sites would be best for the female individuals. The fact that all of the five species investigated by Freeman and his colleagues showed a similar altered ratio of males to females correlated with site characteristics implies that there is a considerable adaptive advantage to this kind of distribution in nature. This would be true regardless of whether the sex differences are genetically fixed or amenable to environmental alteration.

Seeds of ephedras are generally difficult to germinate, although J. A. Young, R. A. Evans, and B. L. Kay succeeded in germinating Nevada ephedra over temperatures with optima ranging from 5 to 20 degrees C. Green ephedra germinated best when exposed to alternating conditions of sixteen hours of cold (2 or 5 degrees C.) and eight hours of warmer temperatures (15 to 25 degrees C.), simulating the conditions it might experience in nature during the spring. Nevada ephedra germinated better than green when exposed to some water stress, in accord with its distribution as more of a desert plant.

Green and Nevada ephedras are of moderate forage value to livestock and deer, and both are considered useless for horses. Frequently, they are of some importance on the winter range. The Indians of the Southwest made extensive use of ephedras. At one time, green ephedra was called *E. antisyphilitica*, in reference to its assumed value in combating various venereal diseases. Nevada ephedra was regarded as similarly useful. It was also variously regarded as a blood purifier, kidney regulator, and cold cure.

There are two varieties of green ephedra, but only one of these, *viridis*, occurs in the Great Basin. The other variety, known as *viscida* or, commonly, as Navajo ephedra, is distributed in the adjacent areas of southeast

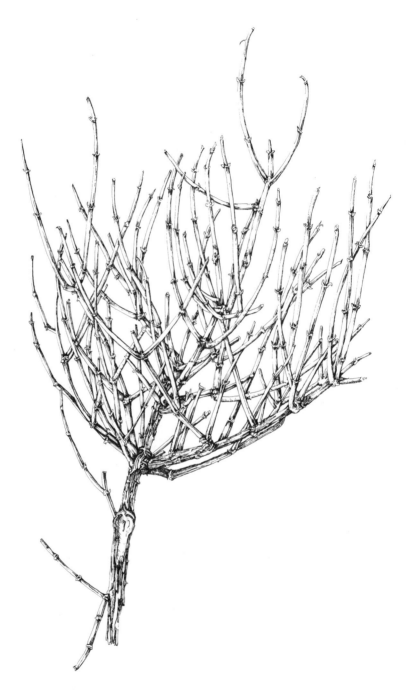

Ephedra, winter aspect

Utah, southwest Colorado, northeast Arizona, and northwest New Mexico. *Viridis* extends into southern Nevada and California, east to the same area occupied by the Navajo ephedra, and north to southwest Wyoming. Similarly, two varieties of Nevada ephedra exist, and, again, only one of these occurs in the Great Basin—variety *nevadensis* extends north into southeast Oregon, west into California, and east into southern Utah and Arizona. A second variety, *aspera*, ranges from southern Nevada down into central Mexico and east to the Rio Grande in Texas.

The genus name *Ephedra* comes from the Greek name for the horsetail plant. There are about fifty species, largely confined to the drier areas of North and South America, the Mediterranean region, and the deserts of Asia. The family Ephedraceae contains only this genus and has no really close relatives. It is probably a very old group and, in a real sense, a relict of the age of the dinosaurs, when it was considerably more widespread. However, *Ephedra* represents a dead end in evolution. We do not now believe that any higher plants evolved from it.

BERBERIDACEAE
BARBERRY FAMILY

Creeping Barberry
Berberis repens

The leaves of the barberry are at times, especially in Europe, infested with a peculiar blight . . . It consists in its full-grown condition of little cups filled with a reddish or brownish powder (spores) . . . This blight caused much fear at one time in Europe, upon the supposition that it was communicated to grain, which however was very probably false. —Charles F. Millspaugh, AMERICAN MEDICINAL PLANTS

UNFORTUNATELY, this opinion, expressed in 1892, turned out to be disastrously wrong—the barberry is indeed the alternate host for the wheat rust, just as the currant is the alternate host for the white pine blister rust. The Pacific Coast species, fortunately, are resistant to this pathogen. Millspaugh's erroneous conclusion is but one more instance of the need for students of nature to be tentative in all their conclusions, however tempting it is to be otherwise!

Creeping barberry, or hollygrape as it is sometimes called, is the most common species of barberry in the West. It is to be expected at higher elevations in coniferous forests throughout the Great Basin. Like those of the other western species, the leaves are composed of leaflets distributed along a common axis, which really represents the midrib of the leaf. Botanists call this a pinnate condition, from the Latin *pinna* which means feather. The creeping barberry usually has about five leaflets making up each pinnate leaf, with each of these leaflets being 3 to 9 centimeters long and spine-toothed along the edge like a holly leaf. The leaves are evergreen and typically occur on short, upright stems only 10 to 30 centimeters tall. Each of these erect stems arises from an underground horizontal stem, or stolon, that gives rise to additional stems at intervals. The individual leaves are dull green above and pale green beneath.

The yellow flowers are borne in a terminal group on the stem and, in common with those of other barberries, have a very regular pattern—consisting of series of whorls of three members each. There are various interpretations of the true nature of each of these whorls but no agreement. One interpretation is that the lowermost whorl is of three bracts, next come two whorls of three sepals often colored like petals, and next are two whorls of concave petals with two nectar glands at the base. To some, these really represent stamens which have evolved to look like petals and have lost the ability to produce pollen. This is not too farfetched; water lilies, for example, clearly show transitional petaloid stamens. Next, there are two whorls of three stamens each. These are initially pressed against the concave petals. They are sensitive to touch, and a bee attempting to get to the nectar glands will stimulate the stamens to spring toward the center of the flower, dusting its head and thorax with pollen. When the bee visits the next flower, the pollen is brushed off onto the receptive portion of the pistil. If cross-pollination does not occur, the anthers eventually contact the pistil and effect self-pollination. Frequently, however, self-pollination does not result in any fruit production. Finally, at the center of the flower is a single pistil, which eventually produces a waxy blue berry 3 to 6 millimeters long. These berries somewhat resemble the fruits of a grape, accounting for the common name of hollygrape.

Somewhat erroneously, we sometimes tell beginning botany classes that the monocots, which include such families as the grasses, lilies, and orchids, have flowers based on a plan of three, while the dicots, which include all other flowering plants, have flowers based on a plan of five. The barberry is a dicot but, however, has the floral plan of a monocot. This does not mean that it is more closely related to the monocots, only that our generalization is false in many instances.

Barberries typically have roots with a bright yellow color on the inside. This is due to the presence of a yellow compound called berberine. The Navajo used the roots for a yellow dye and the fruits to produce a lavender color. Great Basin tribes also used the yellow tea obtained by boiling the roots to thicken the blood and to cure dysentery. The tea was considered

Creeping Barberry

efficacious for curing coughs, kidney problems, and venereal diseases. This and other species of barberry produce edible berries used variously in making preserves, drinks, and pies.

The creeping barberry extends north to British Columbia, east to the Dakotas, and south to Texas and New Mexico. Outside of and adjacent to the Great Basin there are perhaps a dozen species of barberry, a few in the moist coniferous forests of the Northwest, some occurring as large shrubs throughout the warm deserts of the Southwest.

The genus name *Berberis* comes from an Arabic name for the plant, *berberys*, while the species name *repens* comes from the Latin, meaning to creep. In some manuals our species are placed in the genus *Mahonia*. There are about one hundred species of barberry, distributed worldwide. It is the largest genus in the family Berberidaceae, which contains between fifteen and twenty genera and over five hundred species. The greatest number of species occur in cold and temperate areas.

FAGACEAE
BEECH & OAK FAMILY

Bush Chinquapin
Castanopsis sempervirens

OUR CHINQUAPIN GETS its common name from the close resemblance of its fruit burs to those of the eastern chinquapin, *Castanea pumila*. It is, in fact, a close relative, since both belong to the oak or beech family, Fagaceae. The bush chinquapin just barely gets into the Great Basin on the west, along the eastern slopes of the Sierra Nevada and in southern Oregon. It is common on rocky ridges and in open areas in coniferous forests above 6,000 feet. Occasionally, dense thickets are formed, and sometimes it grows mixed with other Sierran shrubs such as tobacco brush and manzanita. On the western slopes of the Sierras, it varies from half a meter to a meter high; rarely, it reaches a height of nearly three meters.

    The oblong, evergreen leaves are blunt and between 2.5 and 8 centimeters long. They have a distinctive rusty or yellowish pubescence on their undersides, which makes this shrub easy to recognize even if there are no fruits present. Male and female flowers are borne in separate clusters on the same plant. Both are minute and grouped in clusters of three. The whitish male flowers are arranged on leafless, catkinlike, but erect rather than drooping stems. These are developed from the base of leaves produced in the same season, and the pistillate flowers are formed along the lower part of these same stems. Blooming starts in early summer and sometimes continues until winter arrives. After pollination by the wind, the pistillate flowers grow into very spiny fruit burs, similar to those of the chestnut. However, maturity does not come until the autumn of the following year. These burs split open by four divisions to release the shiny, yellowish brown nuts, which are sweet and edible, either raw or roasted.

    The bush chinquapin is apparently little browsed by wildlife or livestock, and range managers consider it an indication of overgrazing. It is able to

Bush Chinquapin

stump-sprout immediately after a fire, in the fashion of some manzanitas. The common name chinquapin is an American Indian name.

Ranging from California to Washington is a tree in the same genus, *Castanopsis*. This particular chinquapin, which has the species name of *chrysophylla*, sometimes becomes a tree 30 meters tall with a trunk diameter of 2 meters. This is the only other species of the genus found in North America. Across the Pacific, however, there are about twenty-eight additional species of *Castanopsis*, distributed in Asia from the eastern Himalayas to southern China and Malaysia. This is an example of a disjunct distribution (along with such things as the magnolia and the tulip tree) that botanists believe to be an indication of the close affinity of the flora of eastern Asia and North America, particularly the eastern portion of the latter.

The genus name comes from two Greek words, *castanea*, chestnut, and *opsis*, resemblance. The species name *sempervirens* is derived from the Latin words for evergreen. The Fagaceae family contains about eight genera and around one thousand species, found in both temperate and tropical forests in the northern hemisphere, though one genus, the southern beech, is found in the southern Andes, eastern Australia, and New Zealand. The fossil record of the oak family goes back at least 90 million years to the middle of that period known as the Cretaceous by geologists. This was the period that saw the decline of the dinosaurs and the rapid evolution of flowering plants.

BETULACEAE
BIRCH FAMILY

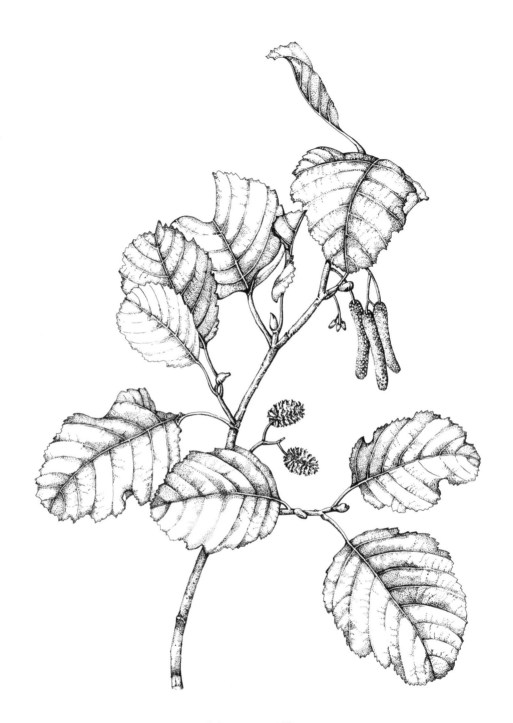

Mountain Alder

Mountain Alder
Alnus tenuifolia

But the Mountain Alder must have its roots in eternal streams whose waters it helps, with its shade and its humus, to keep cool and pure. So this Alder passed into the true legends of the West as the friend of the explorer and settler. —Donald Culross Peattie, A Natural History of Western Trees

Like most of its cousins the world over, our mountain alder frequents waterways in the mountains throughout the Great Basin. Its roots help bind the banks together. Its shade provides a comfortable habitat for those ferns and wildflowers which prefer a moist environment. Ordinarily the mountain alder varies between 2 and 3 meters high, but on rare occasions it may become 8 meters tall. The trunk is smooth and gray or sometimes reddish brown. The dark green, rather coarse leaves are oval to nearly round and from 4 to 10 centimeters long. At first glance, they look a little like elm leaves, but they lack the oblique base of elm leaves and, in the case of our mountain alder, they are edged with both large and small teeth. There is a prominent midrib, and the lateral veins which run at an angle out to the larger teeth are nearly parallel with one another. The leaves begin to drop with the first frosts, frequently without even turning brown.

Alders are easy to recognize even in the winter, or perhaps we should say especially in the winter, because of the clusters of three to seven small, pineconelike structures at the ends of some of the smaller branches. Each "cone," about a centimeter long, consists of the hard, woody bracts which enclosed the winged fruits before they were shed. In addition to last summer's empty cones, elongated staminate catkins, several centimeters long and scheduled to open next spring, are also very obvious at the ends of those branches which developed during the past summer. These branches have characteristic elliptical, reddish brown buds about 6 millimeters long. If one looks carefully, near the base of the dormant staminate catkins will be seen several immature female cones only 3 or 4 millimeters long. It will also be evident that the empty cones from last summer were produced on branches developed during the previous summer. These are located at the base of the most recent growth, and sometimes the cones from two summers ago may still be seen. Last summer's staminate cones, unlike the female cones, are discarded soon after the pollen is released.

The staminate catkins have very small, stamen-bearing flowers borne at the base of the bracts making up the catkin. Each of these tiny flowers consists of four sepals and three or four stamens but no petals. The female cones have two pistillate flowers at the base of each scale; each pistillate flower has a single pistil but no sepals or petals. The flowers are wind-pollinated, and, as is typical of such plants, an overabundance of pollen is produced. The alder is often first to come into bloom in spring, frequently as early as February, if there have been a few warm days. Subsequent to blooming, the leaves rapidly enlarge and reach maturity. Each pistil matures to produce a single small nutlet bordered by a thin wing which aids its dispersal by wind or water.

One notable feature of alders is their ability to fix atmospheric nitrogen. Most people are familiar with this ability as a characteristic of the bean family. In the latter, specialized nodules which contain a certain bacterium occur on the roots. The bacteria are able to trap free nitrogen from the air and assimilate it into living protoplasm. Eventually, this nitrogen becomes available to the host plant. This same bacterium, incidentally, is unable to capture nitrogen when it occurs as a free-living form in the soil. Thus, the bacterium gets nourishment from the host plant, while the host in turn benefits from the nitrogen captured by the "parasite" on the roots. This kind of mutually dependent cooperation goes by the name of symbiosis among biologists.

At any rate, nodules, like those in the bean family, occur on the roots of alders. But these nodules contain a fungus known as an actinomycete, rather than the bacterium common to members of the bean family. This fungus, like the bacterium, is able to assimilate atmospheric nitrogen. Alders, then, aside from the other benefits they confer on their habitats, help

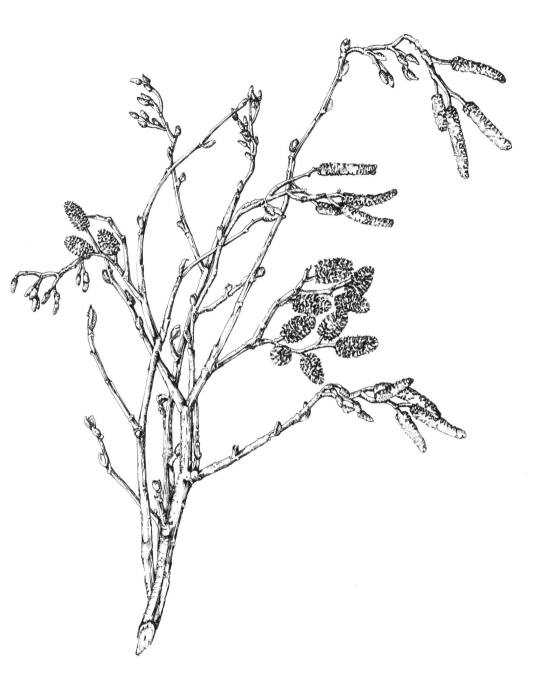

Mountain Alder, winter aspect

add nitrogen to the forest soil. This may be one of the main reasons why the red alder of the Pacific Northwest is the most productive temperate-zone tree, at least according to some authorities. This species of alder has been estimated to yield between 15 and 25 tons of dry matter per hectare per year without any help from humans. As a consequence of the relatively recent appreciation of the role of alders in forest ecology, there have been successful experiments which involved rotating alders with other tree crops or using them as nurse trees. R. F. Tarrant and J. M. Trappe have advocated the use of the red alder either as a rotated "crop," just as members of the bean family, such as alfalfa, are rotated with wheat, or as a member of a mixed planting with a conifer such as the Douglas fir.

D. S. De Bell and M. A. Radwan recently studied the productivity of mixed stands of black cottonwood and red alder. They found that the soil in such plantings showed a 9 percent increase in nitrogen compared to that in pure cottonwood stands. Pure alder stands had 23 percent more nitrogen in the soil. It seems that the old idea of simply planting a pure stand of whatever kind of timber one wants, be it Douglas fir or yellow pine, may not be the best way to get the optimum productivity from forest lands. The diversity that is the pattern for our native forests may be a model of the best way for humans to manage things, after all. Of course, we cannot and do not necessarily want to duplicate this diversity in all its aspects, but certainly we can learn from it.

The alder is of some limited value as browse for livestock and wildlife. Deer sometimes make extensive use of it, especially the younger branches, during severe winters. Other species of alder have been used by tanners, and an extract made from the leaves or bark has been used as a tonic and as an ingredient in bitters. Several of the larger species, notably the red alder of the Northwest, furnish a valuable wood sometimes stained to resemble mahogany by furniture makers. Some alders have been much used for making charcoal. Altogether there are about thirty-five species of alder distributed throughout North America, Europe, and Asia. Several species extend the range of the genus south through Central America to the Andes.

The alders belong to the birch family, the Betulaceae, which contains about 6 genera and over 150 species distributed over the northern hemisphere. Other common representatives are the hazelnut, hornbeam, and hop hornbeam. The genus name *Alnus* means alder in Latin, while the species name *tenuifolia* means slender or thin leaf, not a really appropriate name for this species. Some botanists consider the mountain alder to be a subspecies of the European speckled alder, *Alnus incana,* although most of our manuals do not so list it. The family name is taken from the Latin for birch. The very simple flowers of the alders and birches actually indicate a relatively advanced evolutionary position. This is because many botanists now think that the first flowering plants had large flowers with many petals, sepals, stamens, and pistils that were insect-pollinated. In the case of the birches, the flowers have become much simpler by the loss of parts, a reversion to wind pollination, and a compaction of the inflorescence to form the catkins and cones. The closest relatives of the alders and birches are the oaks and beeches in the family Fagaceae.

CHENOPODIACEAE
GOOSEFOOT FAMILY

Iodine Bush
Allenrolfea occidentalis

ALSO KNOWN AS bush pickleweed and Kern greasewood, the iodine bush gets its name from the dark brown stain left by crushed stems. It is a diffusely branched, erect shrub between 30 centimeters and 1 meter high. The branchlets are green and jointed, like a row of miniature, succulent pickles. The leaves are reduced to mere triangular scales. Iodine bush is found on extremely alkaline or saline flats, frequently with its roots submerged when the dense clay is covered with water during wet springs. Quite often, it will be the only plant visible on such flats. Extensive stands of iodine bush can be seen in the desolate saline areas west of Salt Lake City, where it usually occupies low hummocks on the salt flats. *Allenrolfea* is directly responsible for the formation of these hummocks by capturing windblown sand. Iodine bush produces underground runners which assist in its spread.

The salt tolerance of iodine bush is exceeded by only a few other plants in the Great Basin, such as glasswort or pickleweed (*Salicornia*), an herbaceous species that is able to grow in soil with a salinity as high as 6 percent in the Great Salt Lake region. Iodine bush in the same area can tolerate a salinity of 3 percent but does best at about 1 percent. Seville Flowers found that iodine bush was distributed in a zone behind that of glasswort which extended furthest out onto the salt flats. Some salt-tolerant plants, such as shadscale and tamarisk, are able to excrete excess salt by means of salt glands on the leaves. It is thought that forms such as iodine bush solve the problem of too much salt by storing water and thus becoming succulent, thereby diluting the salt concentration.

Unlike most of the other shrubby members of the goosefoot family in the Great Basin, iodine bush has flowers with both stamens and pistils. On some plants, however, in addition to flowers with both essential components,

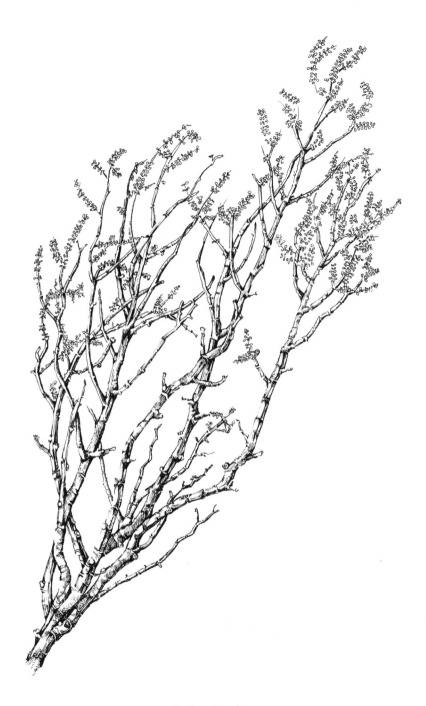

Iodine Bush

there will be flowers with only stamens or pistils. The flowers are formed in scaly spikes at the ends of the branchlets. Each small, inconspicuous flower consists of four or five sepals, no petals, one or two stamens, and one pistil. The flowers are wind-pollinated, and in some areas so much pollen is produced that hay fever may be the result in susceptible persons. The sepals, which become fleshy as the fruits mature, surround the pistil, which contains the single seed.

The range of iodine bush includes California, east to Colorado, New Mexico, and western Texas, and south into northern Mexico. It is abundant on the alkaline flats of southern and northern Nevada and extends northward to eastern Oregon.

The genus is named for Allen Rolfe, a botanist at Kew Gardens in London at the turn of the century. There are two other species in the genus, but *occidentalis* is the only one found in the United States.

Four-winged Saltbush
Atriplex canescens

THE FOUR-WINGED SALTBUSH is the most widely distributed and abundant saltbush in the Southwest. It is found from South Dakota to Texas and west to the arid portions of California, Nevada, Utah, Wyoming, and Washington. In addition, it extends down into Mexico and Baja California. It is found from sea level in California to 8,500 feet in the mountains of Wyoming.

Typically, four-winged saltbush grows on sandy soils and can even be found in sand dune areas, although, not infrequently, it may be found in denser soils more characteristic of greasewood habitat. Larger than the shadscale, it ranges between 1 and 2 meters in height. It is evergreen, and the narrow, lance-shaped, gray-green leaves are 2 to 5 centimeters long, unlike the short, stubby leaves of shadscale. The rigid, brittle main stems are gray, while the younger stems are whitish. Young stems and leaves are covered with minute white scales, a condition known as scurf, which often appears as a kind of plant dandruff under the hand lens. Scurf, found in many desert plants, helps protect against excessive water loss. As might be expected of a plant with such a wide distribution, a good deal of variation occurs in four-winged saltbush, and many of these diverse forms have been named as subspecies, although only two of these are generally recognized now, the typical form and *A. c. aptera*.

Like many other saltbush species, four-winged saltbush has male and female flowers on separate plants, but an occasional individual will have both kinds of flowers present. Male flowers are borne in dense spikes and are each very small, with five sepals fused together at the base into a cup. There are generally five stamens in the interior of the cup. Staminate flowers are borne in clusters up to 15 centimeters long at the ends of smaller branches. The

Four-winged Saltbush

female flower lacks sepals, and there are two small, united bracts which enclose the pistil. These bracts enlarge as the fruits develop, and each bract grows two broad wings, thus giving rise to the common name of four-winged saltbush. Pistillate flowers are born in long spikes which may be 20 centimeters or more in length. Flowering begins in May and can continue until the middle of summer, with fruits maturing in the fall.

E. Durant McArthur, studying the sexual expression of four-winged saltbush over a period of four years, found that some of the plants varied from year to year. There appeared to be three plant types with regard to flower production. One type produced only female flowers year after year, while the second type consistently produced only male flowers. A third type, however, in successive years might change from male to female or the reverse, or, alternatively, might change from a male or a female plant to a plant bearing both types of flowers, or the other way around!

The explanation for this strange state of affairs is a little involved, but, perhaps, not too much so if I leave out some of the extraneous details. First of all, plants which bear only male or female flowers are known as dioecious, whereas those bearing both flower types are called monoecious. Second, in many plants, just as in animals, sexual expression is determined by the presence or absence of certain chromosomes, those structures within the cell that contain the genes or hereditary units. These chromosomes in higher plants and animals come in matched pairs, except for the so-called sex chromosomes. In humans, for example, there are twenty-three pairs of chromosomes. One pair, however, are the sex chromosomes. In males, these two are unlike in size, one being large and called the X chromosome, while the other is small and designated the Y chromosome. In females, the sex chromosomes consist of two similar X chromosomes. This same chromosomal situation with regard to sex determination is known to exist in many dioecious plants. Males are designated as XY, females as XX. However, this explanation is complicated in the four-winged saltbush because of a doubling of the basic chromosome number. All the populations studied, except one, are known to have four sets of chromosomes, for a total of thirty-six. Therefore, McArthur has proposed that male plants in four-winged saltbush have the combination XXYY and that female plants are XXXX. Those plants which are monoecious would be XXXY, and their sexual expression would change depending on environmental conditions. From studies on other organisms, it seems probable that no individuals would be XYYY or YYYY. Probably these are not viable forms in the case of the four-winged saltbush.

One population studied by McArthur, from 1972 to 1973, showed 420 plants with no change and 173 plants that changed sex in one way or another. These figures are consistent with the hypothesis which McArthur has proposed. Another interesting point here has to do with the evolution of separate sexes. Although most higher plants are not dioecious, those that are have independently evolved a chromosomal mechanism for sex determination identical to that of higher animals. And in some plants, as in some animals, it is the female rather than the male that has the two different sex chromosomes. Certain wild species of strawberry are good examples of this latter condition. The mechanism of sex determination in plants is but one more example of the underlying unity of all life, despite the obvious differences in form upon which we sometimes too exclusively focus.

Howard C. Stutz, a geneticist at Brigham Young University, has published many papers dealing with evolution among the saltbushes, and much of what we know today about the natural history of this group can be attributed to his insight. Stutz and his coworkers, J. M. Melby and G. K. Livingston, found a very unusual population of four-winged saltbush in the Little Sahara Sand Dune area in central Utah. Plants in this population were much taller than usual, typically 3 or 4 meters. Fruits were also larger, shoots grew much more rapidly, and seed germination was more successful. Stutz grew plants of this giant or gigas population in the same environment with "normal" four-winged saltbush from other areas and found that these unusual characteristics persisted. It turns out that the gigas population has only two sets of chromosomes rather than the four sets which are typical of all other populations of this species. The gigas population is known as diploid, while the normal populations are termed tetraploid.

Stutz speculates that the rapid growth rates of gigas plants enables them to survive in the shifting sand dunes where they are found. These plants are limited to the leeward side of the dunes, where more moisture would be available because of the greater percolation of water into these steeper slopes. Gigas plants have deep root systems and, after the sand dune has moved on, the newly exposed stems and even the roots develop chlorophyll. The stems and roots also show a strong tendency to develop adventitious roots, another obvious adaptation to a moving dune environment.

Stutz believes that these gigas plants represent a relict population from the time when all four-winged saltbushes were diploid. This doesn't mean that the ancestors of our present four-winged saltbush were giant shrubs—probably most of the specific characteristics which fit the gigas types for a sand dune habitat were secondarily acquired; that is, they are there because of the force of natural selection. Why all except this one population of diploids have disappeared within the last four thousand to five thousand years, Stutz speculates, may be due to the fact that the slower-growing, hardier tetraploid was better adapted to changing climatic conditions and increased grazing pressures. It is very unusual for a dioecious species to give rise to a form with more than two sets of chromosomes, mainly because the separation of the sexes on two different plants means that the offspring of a single tetraploid plant can result only from a cross with a diploid individual. This would result in progeny with three sets of chromosomes, and these are nearly always sterile, for reasons beyond the scope of this discussion.

With the recession of Lake Bonneville and Lake Lahontan following the last ice age, Stutz believes, many new habitats were formed. *Atriplex*, because of its tendency to vary and hybridize, readily provided a host of new variants, many of which were suited to the new environment. These variants differ not only in form but also in physiology. James Young, of the Agricultural Research Service, collected seeds of four-winged saltbush from three different elevations—900 meters, 1,200 meters, and 1,800 meters. Seedlings grown from these collections were exposed to freezing conditions near Reno, Nevada. All the plants from the 900-meters collection died, 6 percent of those from 1,200 meters survived, and 100 percent of those from 1,800 meters survived. This is clear evidence of a genetic difference, regardless of any difference that might exist in form. Plants of the same species which are genetically distinguishable in this way by adaptive characteristics are called ecotypes.

At any rate, Stutz has concluded that an explosive evolution of new saltbush species has occurred in geologically recent times. The way in which the Great Basin is divided into long valleys separated by parallel mountain ranges has resulted in smaller populations of saltbushes, effectively isolated from their neighbors in the next valley. As we will discuss later, small populations of this sort would be expected to undergo a much more rapid pace of evolution than larger, not so isolated populations. Stutz remarks that virtually every known strategy for the evolution of new species is presently operating in the saltbushes.

Craig A. Hanson, who studied the perennial species of the genus *Atriplex* of Utah and the northern deserts for his master's degree, found seventeen species in the area he surveyed. There are also numerous hybrids and subspecies, some of which he found and described for the first time—as a result of his work, he described three new species and one new subspecies. Of the total of seventeen, eleven are classified as herbs or subshrubs rather than true shrubs, which are woody throughout. One interesting observation was that, while the completely shrubby forms occasionally hybridize with one another, they very frequently cross with the subshrubby and herbaceous forms. Stutz suggests that, since woody and herbaceous species hybridize readily and produce offspring which are at least partly fertile, the hereditary basis for the production of wood must be rather simple.

As a forage plant, four-winged saltbush is highly esteemed. Cattle readily eat the leaves, stems, flowers, and fruits, and the fact that it is evergreen makes it valuable on the winter range. As might be expected, it is also an important cover and food source for a variety of wildlife. Shrubby associates of four-winged saltbush commonly include other species of saltbush, particularly shadscale, as well as Bailey's greasewood, winterfat, and spiny hopsage. On occasion four-winged saltbush is the dominant species over a large area. Aside from the typical form, another subspecies is presently recognized; however, it is not found in the Great Basin.

The species name *canescens* comes from the Latin and is a reference to this plant's whitish appearance. Other common names applied to it are saltsage, wafer sagebush, buckwheat shrub, and chamiza. The name four-winged, which is perhaps most appropriate, was suggested by Frederick Coville, famous for introducing the blueberry into cultivation.

Shadscale
Atriplex confertifolia

THREE SHRUBS ARE especially characteristic of the Great Basin—sagebrush, rabbitbrush, and shadscale. Of these three, shadscale is capable of growing in the driest locations and is truly a desert plant. I really believe that Mark Twain's characterization of the sagebrush would have been better applied to the shadscale, though perhaps "the fag end of vegetable creation" is too harsh an appellation even for the shadscale. For, as nondescript as the shadscale may at first appear to be, it belongs to a very successful plant family, one that is quite advanced in several respects.

The generic name *Atriplex* was the old Latin name for this group of plants. They are especially common in deserts around the world, particularly those with saline or alkaline areas. In many countries, members of this genus are referred to as saltbushes. The family to which shadscale belongs is distributed from southern South America, Australia, and South Africa to North America, Europe, and Asia. It is especially prevalent around the shores of the Mediterranean and in the deserts and steppes of central and eastern Asia. In our own Great Basin Desert, various members of this family are invariably the ones able to tolerate, more than any other higher plant species here, the extremely saline conditions at the edges of our playas. But it is also obvious that even the saltbushes have not yet been able to evolve a way of coping with the extreme dryness and salinity characteristic of the vast sterile areas of many of our playas.

Other common names for the shadscale are spiny saltbush and saltsage. The specific name *confertifolia* means a dense cluster of leaves. In reality, this refers more to the clusters of fruiting bracts than to the leaves. The scientific name for the family which includes the shadscale is Chenopodiaceae, derived from a Greek word meaning goose foot, an allusion to the leaf shape

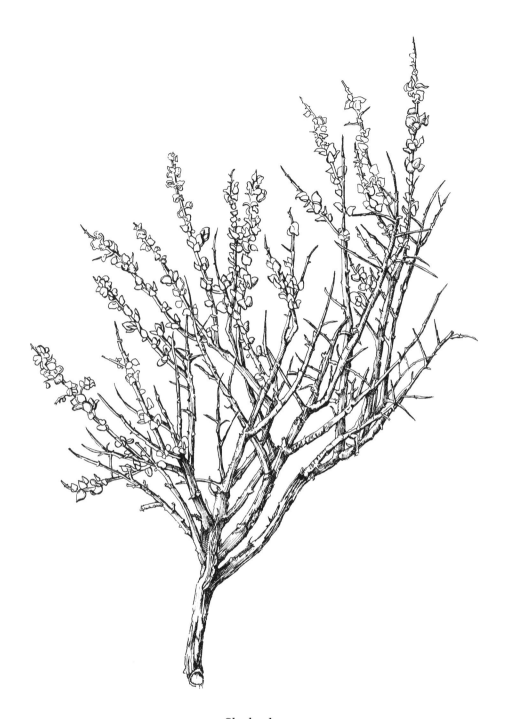

Shadscale

of many of its members. For example, the common weed called lamb's-quarters, known botanically as *Chenopodium album*, shows the leaf shape after which the family is named. Aside from the important role played by this family in nature, it has provided some important cultivated vegetables, notably the beet and its variety the sugar beet, Swiss chard, and spinach.

Shadscale bushes will occasionally reach a meter in height, although typical examples are only a third as tall. Usually they are rather rigidly branched, spiny, and somewhat compact rounded shrubs. Grazing seems to make them even more compact, and there is some evidence that they are benefited by a moderate amount of use by cattle. Although it is generally considered less palatable than black sagebrush, shadscale is so abundant on winter ranges that it is regarded as a very important forage plant. Both the leaves and the seeds are eaten by cattle, as well as the branchlets before they become hard. Sheep also browse the herbage and even lick fruiting bracts up off the ground. The small branches bear leaves at first but taper to a point and become woody, hard spines after the leaves are shed. They afford the plant some protection from grazing and, not surprisingly, in areas which are too heavily grazed, they have been known to cause mouth problems. The leaves are generally somewhat ovate in shape and up to 2 centimeters in length. They are shed during the fall, and new leaves are not produced until March or April. The leaves are gray or somewhat tinged with red and scaly in appearance. The young branchlets have a straw color.

Male and female flower parts are always produced on separate plants. This is unlike the arrangement in most plants, in which both the stamens and the pistils occur together in the same flower structure. There is a basic adaptive advantage to the occurrence of the two sexes on separate plants—it represents, in a real sense, the logical extreme of one mode of evolution. Put very simply, the more the variable traits that exist, the more probable it is that at least some portion of the population will be preadapted to any change that comes along. To this end, many flowering plants have evolved a variety of techniques to insure that pollen will not accomplish the fertilization of the pistils on the same plant. In many plants this involves as simple a procedure as the release of pollen either before or after the pistil becomes receptive. In certain advanced types, such as the orchids, which are insect-pollinated, the arrangement of the stamens and the receptive portion of the pistil (the stigma) makes self-pollination impossible.

An alternate direction that evolution has taken in this regard is seen in cases where the stamens are borne in flowers separate from the pistils, but still on the same plant. Corn is a good example of this pattern, for the ear has only female flowers while the tassel contains only male flowers. When the male and female parts are in separate flowers but on the same plant, we call this condition monoecious. The word comes originally from the Greek and means, literally, one who dwells alone. The condition which occurs in the shadscale and many other members of the same family in which the two kinds of flowers are borne on separate plants is known as dioecious, which means dwelling in two houses.

Most people today are familiar with the fact that there are two human chromosomes which determine sex. In humans these are called X and Y chromosomes. All females normally contain two X chromosomes in each of their cells—represented symbolically as XX. All males contain only one X chromosome and another, smaller one called a Y chromosome. In some way not yet understood, this chromosomal difference is responsible for the determination of primary and secondary sexual characters. In most plants there are no obvious sexual chromosomes, but in many dioecious forms they have been identified. That is, the female plants have the sex chromosome complement XX, while the males are XY. These sex chromosomes are in addition to the other pairs of chromosomes always present.

When reproduction occurs, the two sex chromosomes are separated, so that all the egg cells have only one X chromosome; the males, on the other hand, produce two kinds of pollen grains, one containing a single X chromosome while the other contains the Y chromosome unaccompanied by any X chromosome. Perhaps the most interesting aspect of this particular mechanism is the fact that a similar chromosomal mechanism for sex determination has evolved independently in both plants and animals. Similar instances of convergent or parallel evolution have occurred many times in plants. The succulent cacti, for example, resemble the entirely unrelated succulent euphorbias of Africa. And spurred flowers have evolved independently in violets, delphiniums, and orchids. Charles Darwin, in his famous study on the finches of the Galápagos islands, was the first to describe an adaptive pattern that resulted in a convergence or parallel evolution of similar forms.

So far, no one has definitely identified sex chromosomes in the shadscale, although there is fairly good evidence that they occur in the four-winged saltbush and by implication in the shadscale as well.

The staminate flower in the shadscale is quite small, consisting of only three to five stamens, but many are grouped together in spherical clusters on short, lateral branchlets. The pistillate flowers contain a single pistil enclosed by two leafy bracts fused together at the base. Typically, several pistillate flowers are borne near the ends of smaller branches at the base of the foliage leaves. Since the flowers are relatively inconspicuous, it would be a good assumption that they are wind-pollinated, as is the case with the sagebrush. Wind pollination would seem to be such a hit-or-miss affair that it comes as somewhat of a surprise to learn that certain very successful plant groups, both primitive and advanced, are dependent on it. All the conifers—pines, firs, hemlocks, spruces, etc.—are wind-pollinated, as are the relatively advanced poplars, birches, and oaks. There is some inefficiency, of course, in wind pollination. Most of the pollen never arrives at the female flower, and we find, typically, that a tremendous excess of pollen is produced.

On the other hand, wind-pollinated plants are not dependent on insects for seed production. Obviously, in the case of insect-pollinated plants, there is a greater efficiency in getting the pollen to the right place. The major disadvantage, if it can be called such, is that the survival and evolution of insect-pollinated forms are inextricably bound to those of the pollinating insects. The two must evolve together, more or less mutually, if they are both to survive. This process is sometimes called coevolution.

In any case, seed production is fairly successful in the shadscale. Juan M. Gasto carried out lengthy studies of the development, growth, and ecological relationships of shadscale and winterfat shrubs in the wild in Curlew Valley, Utah, and in the greenhouse. He found that shadscale, during the season he measured it, produced an average of 2,204 fruits per plant. Only about 18 percent of the fruits contained seeds, however. For some unknown reason, there was better seed production in areas with mixed stands of the two shrubs than in areas with shadscale alone. In mixed stands, shadscale fruits contained seeds 25 percent of the time. Curiously enough, this difference persisted when it came to germination percentages. Germination of shadscale seeds in soil taken from mixed stands was significantly better than that in soil from pure shadscale stands. And it was also better in soil from the upper 2.5 centimeters than in soil from below this level. Apparently a chemical difference in the surface soil somehow encourages germination.

Gasto also looked at the amount of growth of shadscale seedlings in various soils and found that they grew best in the surface soil from pure stands of shadscale as well as in that from another species of saltbush. In fact, growth in soil from a pure winterfat area was nearly as good. The important conclusion that Gasto arrived at was that, although soil is important, in this case soil differences cannot be used to explain the presence or absence of shadscale in a given area. But there are population differences, nevertheless, which must have some causal basis. In cases such as this, some ecologists invoke a hypothesis known as Gause's rule, which dogmatically states that no two species can occupy precisely the same niche in nature. If the two species have essentially the same requirements, then, by competition, the population of one species will increase until the other species is eliminated. There is a good deal of debate about this idea, however, and it is a very difficult concept to demonstrate to everyone's satisfaction. Complicating this whole picture are the results of other workers, who have shown that shadscale seedlings from different locations generally do better in their own soil than in soil from other shadscale areas. This implies that the shadscale populations are made up of a number of genetically distinct ecotypes, each adapted to its own specific location. Such ecotypes would be genetically different from one another.

Migration is another factor that determines which plants will be found where. Gasto found that, for each 4,093 fruits of winterfat migrating, only 4 will remain alive at the end of the first year and only 1 at the end of the third year. For shadscale, of 2,353 fruits migrating, only about 5 seedlings will be alive at the end of the first year, only 1 at the end of the third year. Shadscale fruits, because of their weight, are not easily dispersed from the parent plant, and Gasto concluded that dissemination and migration are the limiting factors and the reasons why there are not more seedlings of winterfat or shadscale in pure stands of the other species. In short, even though shadscale can grow in more places than it is found, it first has to get there in sizable numbers in order to prevail.

Dwight Billings, some years ago, studied the plant communities of the western Great Basin and was able to recognize some fifteen different associations in the Carson Desert region. In two of these, shadscale plays a major role. In the little greasewood–shadscale association, the Bailey's or little greasewood (*Sarcobatus baileyi*) shares a dominant status with shadscale. This association occupies more area in the Carson Desert than any other. Billings found the soils on which it occurred to be gravelly and generally well drained. This is the typical association in those areas with so-called desert pavement, which are the result of wind erosion exposing a somewhat uniform gravelly surface. The dark and shiny appearance of these rocks is due to the development of a coating of manganese dioxide on the surface rocks under desert conditions. Although Billings did not find this to be the case in the Carson Desert region, in other places shadscale may be the only dominant shrub, and little greasewood will be absent.

Arthur Cronquist, Arthur Holmgren, Noel Holmgren, and James Reveal, in their monumental six-volume *Intermountain Flora*, which is still in progress, refer to this association as the shadscale zone. Along with Billings, they regard shadscale's presence in desert areas as being due to its tolerance of dryness as much as its resistance to saline conditions. Gasto reached much the same conclusion and added that both winterfat and shadscale were physiologically able to grow in a wider variety of environments than they were actually found in.

The famous geneticist G. Ledyard Stebbins has pointed out that the shrubby members of the goosefoot family undoubtedly evolved from herbaceous ancestors, a reversal of the usual direction in evolution. Some 70 million years ago, the goosefoot family was beginning a very active phase of evolution. While at that time the arid portions of North America were small in extent, they began to enlarge over the next 40 million years, with the result that natural selection favored the evolution of shrubby, evaporation-resistant forms. Since these shrubs were derived from nonwoody types, it is not too surprising that their stem anatomy differs significantly from that of the more conventional shrubby types found in more humid climates. Instead of a single layer of cambium-producing wood or xylem on the inside and food-conducting tissue or phloem on the outside, they have a series of concentric layers of cambium, like the layers of an onion, each producing its own phloem and xylem. In the deserts of central Asia, there are even tree forms of the goosefoot family with this same anatomy.

Another interesting aspect of the physiology of some *Atriplex* species is their ability to carry on a special, more efficient kind of photosynthesis in addition to the more common kind. Without going into the biochemical details, we can say that this particular pathway, known as the Hatch-Slack or C_4 pathway after its discoverers, has evolved independently in a number of tropical and arid-zone plants. The high productivity of sugarcane and corn is due to the Hatch-Slack pattern of photosynthesis. This pathway has recently been found to exist in shadscale.

Shadscale ranges from southern Idaho to Wyoming and south through the Great Basin to New Mexico and Mexico. It is considered palatable to all domesticated grazing animals and, as we mentioned, it is generally regarded as an important winter forage plant. In some areas it has been overgrazed despite the prevalent spines; in wet weather the spiny branches become softer and can be grazed to some extent. The fruits and leaves of shadscale provide important food for a variety of other desert denizens, notably many small rodents, jackrabbits, and deer, as well as game birds and song birds.

Torrey Saltbush
Atriplex torreyi

AS THE TALLEST saltbush in the Great Basin, the Torrey occasionally attains heights of over 3 meters. It is distributed from the Mohave Desert and Owens Valley in California, to the Colorado River and Lake Mead area of southern Nevada, northward to extreme southwestern Utah and central and western Nevada. Several prominent stands of Torrey saltbush can be found on the floodplains of the Truckee River near Pyramid Lake. It is absent from most of the Great Basin, however. Like greasewood, it grows in heavy saline or alkaline soils, particularly where subsurface water is readily available. Its associates include greasewood, four-winged saltbush, and desert blite. Occasionally, it can be found in relatively pure but small stands.

The leaves in Torrey saltbush, oval or shaped somewhat like an arrowhead with small lobes at the base, are from 1 to 3 centimeters long. They are whitish or gray-green in color, very much like those of the four-winged saltbush. Aside from the size of the plant and its leaf form, Torrey saltbush is easily distinguishable from the other species in our area by its very small fruiting bracts, which are only 2 to 4 millimeters in length. Like those of the shadscale, these two bracts are fused with the pistil and have no extra "wings" like those of the four-winged saltbush. Both the male and the female flowers are produced in dense clusters at the ends of the branches. Frequently, both sexes are borne on the same plant, while other individuals appear to be either one type or the other. Unfortunately, no one has yet studied this species as thoroughly as McArthur did in the case of the four-winged saltbush, but it is logical to expect that the same condition might prevail in this case—that is, some plants will change their sexual expression depending on the environment.

A close relative of Torrey saltbush is the quail brush or lenscale, *A. len-*

Torrey Saltbush

tiformis. In fact, some authorities today consider the Torrey saltbush as only a variety of the quail brush. The two are easy to separate, however. The Torrey has sharply angled branchlets because of the presence of raised longitudinal ridges, and it is frequently quite spiny, whereas the quail brush has rounded branchlets and tends to be not as spiny. Furthermore, the quail brush does not enter the Great Basin. Its distribution includes the Colorado and Mohave deserts in California and the Colorado and Virgin River areas of Nevada and adjacent Utah. It extends southward well into Baja California and Sonora, Mexico.

Although not abundant enough in the Basin to be significant as forage, Torrey saltbush is palatable to livestock and is important as cover and as a source of food for wildlife.

The species is named after John Torrey, the prominent American botanist of the nineteenth century who described and named many plants brought east by collectors on early expeditions to the western United States.

Saltsage
Atriplex tridentata

SALTSAGE IS A member of a complex of several related species of small stature found in heavy, alkaline, clay soils of valley flats in the eastern Great Basin. It is particularly abundant in the Lake Bonneville Basin in Utah. This entire complex, unlike the shadscale, four-winged, and Torrey saltbushes, is only partly woody or subshrubby. Some individuals, in fact, appear to be wholly herbaceous and die back to the woody rootstock each winter. Saltsage may get to be a meter tall, though it is generally much less. Most of the branches grow upright from the root crown and bear pale gray-green, narrow to oblong leaves between 1 and 5 centimeters in length. The sexes are mostly on separate plants. The female flowers, clustered at the ends of the branches, have two united bracts surrounding the pistil. In the mature fruits, these bracts have three or more prominent teeth at the apex and are smooth or, more frequently, possess bumps or tubercles on the surface. The yellow or brown male flowers are also borne in clusters at the ends of the branches.

Howard C. Stutz of Brigham Young University and his colleagues, C. Lorenzo Pope and Stewart C. Sanderson, have carried out extensive evolutionary studies on saltsage and its close relatives. They found that every population of this species studied is distinctive and unique. One population in Tooele County, Utah, is unusually tall (growing 90 centimeters in one season), is somewhat woody, and has large fruits and leaves, while another population near Delle, Utah, is nearly prostrate and has slender branches as well as small leaves and fruits.

Interestingly enough, most saltsage populations are quite uniform within. Stutz attributes this to two factors: the strong root-sprouting tendency of saltsage, which makes for a lot of vegetative reproduction within each

Saltsage

colony, and a genetic attribute associated with species that have more than the usual two sets of chromosomes, the so-called polyploids. In our discussion on the four-winged saltbush we point out that all populations studied, except one, were found to be tetraploid; that is, they had four sets of chromosomes. In the case of saltsage, Stutz found that most populations are hexaploid—they have six sets of chromosomes for a total of fifty-four. Stutz speculates that the enormous variability between individual populations of saltsage is a consequence of the unexploited habitats exposed in the Great Basin by the recent disappearance of the Ice Age lakes, Bonneville and Lahontan. The desert "islands" separated by parallel mountain ranges have effectively isolated each population, and in such instances, in small colonies, evolution tends to be accelerated as chance variations spread rapidly through the population.

Natural hybrids between saltsage and four-winged saltbush were found by Stutz at a number of sites. These individuals had five sets of chromosomes for a total of forty-five. These hybrids in nature provide, in subsequent generations, a bridge by means of which some genes could be transferred between these two species. This process results in adaptive combinations which may eventually become new species. One of these new combinations is a low-growing form in the Reese River Valley, Nevada; another is a robust form near Knolls, Tooele County, Utah; and one is an upright, bushy type near Grantsville, Tooele County, Utah. All of this evidence makes apparent the truth of Stutz' observation that the saltbushes are in a period of explosive evolution. If so much has happened in a period of a few thousand years or less, one wonders what the saltbushes and their allies will be like in another few hundreds or thousands of years. If humans are still around then, it is comforting to realize that botanists will still have plenty of things to study, what with these new forms constantly being spewed out by the forces of evolution.

Sometimes saltsage grows in nearly pure stands, and it appears to be able to withstand very high concentrations of salt in the soil. Similar to saltsage but less able to withstand an extremely saline environment is the sickle saltsage, *A. falcata*, which can be distinguished by the single, frequently curved, long, pointed tooth at the top of each fruiting bract. Sickle saltsage is not really a shrub, however, since it is woody only at the ground level. Craig A. Hanson, who did a very thorough study of the perennial *Atriplex* species of

Utah and the northern deserts, found that the sickle saltsage was scattered throughout the Great Basin but was nowhere very common.

Perhaps the most significant contribution of Hanson in his study of this group was the recognition of the Bonneville saltsage, *A. bonnevillensis,* as a distinct species. He plotted the distribution of this form through central and northern Nevada and western Utah. Apparently, this species occurs only in the Great Basin. A fair-sized shrub, it may grow to a height of 75 centimeters. It is woody, at least in the lower third, and sometimes entirely so. It somewhat resembles four-winged saltbush and in fact, like it, has four wings on the fruit. However, the fruiting bracts are under 9 millimeters wide, whereas those of the four-wing are larger. In addition, while the tips of the fruiting bracts of four-wing are generally much shorter than its wings, the bracts are longer than the wings in the case of Bonneville saltsage. A color difference also exists in the male flowers—those of four-wing are yellow, while Bonneville has mostly brown ones. The evidence appears to be good that the Bonneville saltsage originated as a consequence of hybridization between four-winged saltbush and sickle saltsage. This apparently happened several times in different locations. It seems reasonable to regard this as a valid species rather than simply another hybrid because it now appears to be stable and, in addition, has different soil requirements than either of its parents. Its leaves are shorter than either of the parental species, and also unlike them, it is able to grow on heavy, saline soils that are periodically flooded. Apparently this saltsage has been collected numerous times since 1865, but the Bonneville was simply not recognized as a separate species until Hanson's insight made its distinctness apparent.

The specific name of the saltsage, *tridentata,* refers to the three teeth at the end of the fruiting bract. In the sickle saltsage, the specific name comes from the Latin *falcatus,* which means scythe- or sickle-shaped, in reference to the terminal tooth on the fruiting bract.

Winterfat
Ceratoides lanata

Scarcely differing from narrow-leaved forms of the Asiatic E. ceratoides. From the Saskatchewan and Western Dakota to New Mexico, and westward to the Sierras. Frequent in the dry valleys and ridges of Nevada and Western Utah, retaining its foliage and fruit through winter, and valuable for its fattening qualities for stock. Beef thus fed, however, acquires a peculiar rather disagreeable flavor. Known both as "White Sage" and "Winter Fat," and of repute as a remedy in intermittents. —Sereno Watson, BOTANY

ALONG WITH SAGEBRUSH, rabbitbrush, bitterbrush, and shadscale, winterfat or whitesage has been the subject of numerous and intensive studies, this because of its considerable positive value for livestock. Dwight Billings in 1945, characterizing the plant associations of the Carson Desert region in western Nevada, noted that the pure stands of winterfat were abundant enough in that area to warrant naming a winterfat association. Actually, winterfat is common all over the Great Basin, frequently in shadscale-greasewood associations and sometimes in pure stands. H. L. Shantz noted in 1925 that it often becomes dominant where shadscale has been killed. Over the Great Basin and extensively beyond it, winterfat is a conspicuous and dominant plant, especially during the winter. In the north it is associated with saltbush, rabbitbrush, sagebrush, and even greasewood. To the south it can be found in mesquite communities.

D. H. Gates, L. A. Stoddart, and C. W. Cook attempted to determine what soil factors might control the distribution of winterfat and four other desert shrubs. Although there were variations, in general, soils under sagebrush were lowest in salts, followed by winterfat, shadscale, greasewood, and

Nuttall's saltbush. They found that with regard to exchangeable sodium, i.e., sodium which could be replaced by other ions in the soil solution, soils under winterfat had the smallest amount. Winterfat soils, on the average, were lightest in texture. Richard E. Eckert studied competition between winterfat and the obnoxious introduced weed halogeton and found that soil temperature had a great deal to do with seedling establishment of both these species. Early in the season, winterfat seedlings showed a rapid root growth, apparently best between 4.5 and 15.5 degrees C. Halogeton roots grew best from 15.5 to 26.7 degrees C. This resulted in winterfat developing a much larger root system until the first week in July, when halogeton roots began to grow more rapidly—by the end of the season, the latter's roots were five times the total root length of winterfat. Eckert's other studies on winterfat led him to conclude that overgrazing by rabbits or livestock during the winter or spring produced plants with reduced vigor and, as a result, halogeton could easily succeed in its invasion of such areas.

In his work on a degree at Utah State University, Juan M. Gasto studied the autecology of winterfat and shadscale. Autecology simply means a study of those ecological factors which influence the development and distribution of a particular plant species. Gasto studied the effect of a variety of soil and climatic variables on seed viability and germination. He found that the germination tolerance for both species was much broader than the magnitude of the soil factors where they naturally occur.

One of the most difficult problems in nature involves trying to understand why a given species grows in one place and not another, particularly when the species appears to be adaptable enough to grow in either location. This was the sort of puzzle that confronted Gasto. Obviously, plants which are physiologically not fitted to survive in a certain environment are not going to survive there. So, one line of investigation involves finding out something about the climatic and soil optima and limits for the growth and development of individual species. However, this kind of measurement of tolerance limits really does not work very well in explaining the distribution of individual species, particularly when it comes to understanding why a species cannot be found in areas which are well within its environmental requisites. As Gasto puts it with regard to shadscale and winterfat, "Their range of physiological tolerance is broader than the magnitude of the actual environmental factors found in the areas where they are present."

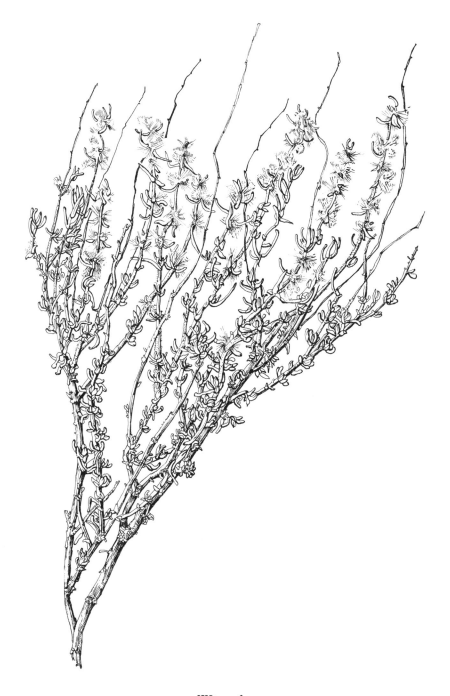

Winterfat

Gasto cogently points out that we can understand this situation only if we remember that we are dealing with populations, not simply individuals. There is a principle of competitive exclusion which states that no two species can occupy exactly the same niche in nature. Which species will be present depends on three factors: reproductive rate, death rate, and migration rate. Gasto did not study the reproductive or death rates in natural communities, but he did observe that migration rates of both shadscale and winterfat were low. To a great extent, then, within the limits of their environmental tolerance, individual species are found in certain areas because of historical accidents of migration and establishment. All of the foregoing implies, perhaps more than we would like to admit, that chance plays a major role in plant distribution and the makeup of individual plant communities.

Russell Moore, also a graduate student at Utah State University, studied transpiration (water evaporation) from shadscale and winterfat in relation to various soil and atmospheric factors. He found that these plants were able to extract sufficient soil water, to carry on photosynthesis and evaporate water from their surfaces into the air, at soil moisture levels which would kill most nondesert plants. As the season progresses the leaves of winterfat become more hairy, and they show a strong tendency to curl lengthwise toward the underside, a condition which botanists call revolute. Both these developments help reduce water loss from the leaves still further. Moore found that the leaves developed from spring buds in both shadscale and winterfat tended to be large and succulent and were only sparsely covered with hairs (or scales in the case of shadscale). But during midsummer these spring leaves died and were gradually replaced by smaller, more compact leaves which were densely covered with hairs or scales. Many other desert shrubs, such as the littleleaf horsebrush, don't attempt to produce any midsummer leaves; by that time, they have finished their growth for the year. This is a strategy similar to that of many desert animals, which go underground during the day and simply avoid trying to do anything when conditions are too hot.

Another study comparing shadscale and winterfat was carried out by M. M. Caldwell and his coworkers in Curlew Valley, Utah. They found that winterfat had curtailed almost all photosynthetic and transpiratory activity by the first part of August, while photosynthesis continued much later in the season for shadscale.

Leaford Windle, a graduate student at the University of Idaho, investigated six perennial species of the salt desert shrub in south central Idaho, among them winterfat. He found that winterfat which was protected in an exclosure from grazing animals increased considerably in vigor over a period of seven years, while those outside the exclosure declined and were in very poor condition. In another exclosure, Windle measured the development of individual plants. He found that red shoot buds began appearing by the middle of March. A few new leaves were present in early April, when the stems were just beginning to grow. Growth was well along by June, and the male flowers began to produce pollen by mid June. Seeds were developing by July, and by the end of that month some of the leaves had died. Seeds appeared not to be mature until mid October.

Winterfat typically is about 30 centimeters high, but it may on occasion reach a height of 1 meter. With a hand lens, star-shaped and simple, unbranched hairs can be seen covering the entire plant. Initially white or grayish, aging hairs turn a rusty color. The leaves are narrow and up to 4.5 centimeters long; frequently, clusters of smaller leaves occur at the stem nodes. Leaf edges are curled toward the underside. Winterfat has a deep taproot and numerous branched lateral roots, which aid in its resistance to drought. Individual flowers are either male or female. Commonly both types occur on the same plant, but sometimes individuals are exclusively male or female. Like those of other members of the goosefoot family, the flowers have no petals and are small and relatively inconspicuous. There are four sepals and four stamens. Female flowers have no sepals but only two fused, hairy, and persistent bracts united around the single pistil. The silky, long hairs surrounding the flowers sometimes appear to be arranged in four dense tufts. Eventually the small, hard fruits, covered with silvery hairs, reach a length of about 5 millimeters.

Winterfat is generally regarded as one of the most desirable winter browses found on subalkaline flats for all domestic livestock. It is an important winter browse for elk and deer. Jim Young reports that, despite its obvious value as a forage and despite fifty years of research, very few stands have ever been artificially established by seeding. Watson's observation about the bitter taste produced in beef by winterfat is not regarded as a credible notion today by range workers—though, of course, it is rare for cattle to be fed exclusively on winterfat. Long ago, J. G. Smith, in his publication on fodder and forage plants for the U.S. Department of Agriculture, thought that livestock fed on winterfat were "remarkably free from disease because of the tonic properties of the plant." Winterfat is a good source of protein and vitamin A. Great Basin Indians boiled the leaves and stems to prepare an extract which they believed to be efficacious in treating eye problems. The hot solution was also used to get rid of head lice and, in fact, was highly regarded as a hair and scalp tonic, even to the extent of restoring hair in cases of baldness!

The range of winterfat extends from Mexico and southern California north to British Columbia and Manitoba and east to the Rocky Mountains, western Texas, and Nebraska.

Winterfat is a member of the goosefoot family, the Chenopodiaceae, some of the characteristics of which are described under green molly and shadscale. For a long time, winterfat was known by the genus name of *Eurotia*. However, John Thomas Howell of the California Academy of Sciences found that the name *Eurotia* was misapplied, according to the rules of botanical nomenclature, and that the correct genus name was *Ceratoides*. This name is derived from the Greek *keratoeides*, meaning like a horn in shape or hardness, in reference to the two small, short horns on top of the fruits. The genus consists of only two species, the other one occurring in the deserts of Asia. The species name *lanata* is from the Latin *lana*, which implies wool or hair.

Spiny Hopsage
Grayia spinosa

THE MOST FAMOUS American botanist of the nineteenth century and one of the most influential protagonists of Darwin's evolutionary theory in this country was Asa Gray. A French botanist, Christian Moquin-Tandon, decided to honor Gray by naming this genus of Great Basin shrubs after him in 1894. The common name of spiny hopsage is particularly appropriate, since its fruiting structures closely resemble those of the hops of commerce. The individual pistillate flowers, which are clustered toward the ends of the branches, are enclosed by a pair of prominent bracts almost completely united, except at the tips. These bracts may eventually get to be over a centimeter wide. At maturity they may be yellow, like hops, or tinged with red. Like those of the other members of the goosefoot family, the female flowers are relatively simple and lack both petals and sepals. The male flowers, which are produced in small clusters at the base of leaves or bracts, have four or five sepals surrounding an equal number of stamens. Some plants seem to produce only male or female flowers, while others produce both types. Pollination, as is also true of the hopsage's relatives, is by means of the wind.

Although many manuals call it an evergreen, spiny hopsage, at least in the Great Basin, loses most of its leaves during the winter. Even during that season it is an easy plant to recognize, since its spine-tipped twigs are reddish and appear to have loose, whitish strings of bark attached. None of our other shrubs has this distinctive characteristic. Small, gray winter buds are also evident. Spiny hopsage leaves are blunt and vary somewhat in shape from narrow to relatively wide. They may be as much as 30 millimeters long under moist conditions. The whole plant is generally a meter or less in height.

The range of the spiny hopsage includes not only all of the Great Basin but southern Nevada and California and eastern Washington and Wyo-

Spiny Hopsage

ming. A distribution such as this indicates a high degree of plasticity in habitat requirements. Spiny hopsage is, in fact, one of the few woody shrubs which is at home in big sagebrush, shadscale, pinyon-juniper, and creosote bush communities. Its altitudinal range is comparably great, from 2,500 to 7,500 feet. Only one other species of hop-sage exists, *G. brandegei*, the spineless hopsage. Its range, largely outside the Basin, extends from southern Wyoming through Colorado and Utah to Arizona. It seems to be localized primarily on clay soils. The spineless hopsage is easily recognized, as the name implies, by its lack of spines. In addition, the fruiting bracts surrounding the seed are notably smaller.

By ranchers, the spiny hopsage is considered to be a fairly valuable browse species, so much so that a number of experiments have been conducted on seed germination and propagation by means of cuttings. The younger portions are eaten, including the fruiting clusters. After the leaves and fruits have fallen during autumn and subsequently collect in depressions, they will be consumed by sheep. Arthur W. Sampson and Beryl S. Jesperson, in their comprehensive work on California range brushlands and browse plants, consider the spiny hopsage as good to fair for sheep, goats, and deer, fair to poor for cattle, and poor to useless for horses.

Grayia has the ability to accumulate certain elements in its leaves, with the result that the surface soil layers under the spiny hopsage will frequently show high concentrations of potassium and magnesium. Unlike some of its relatives in the goosefoot family, it does not, however, have the ability to grow in saline soils and seems to be confined primarily to alkaline areas. It will grow in dry as well as wet sites, on mesas and flats, and on rocky talus slopes and within steep-walled canyons. An indication of its genetic variability, aside from its wide distribution, was apparent in the results of studies undertaken by M. K. Wood, R. W. Knight, and J. A. Young. They found, interestingly enough, that hopsage seed from the Mohave Desert germinated at 40 degrees C. but that seed from Nevada would not. The best germination, however, occurred with an alternating temperature regime of cold and warmth, reflecting the natural cycle expected in desert climates. They also found that only a little moisture was required to initiate germination and that the bracts in some unknown way helped germination when there was some water stress, such as might be produced by dissolved salts in the soil.

Another experiment, initiated by Burgess L. Kay of the Department of Agronomy and Range Science at the University of California at Davis, was intended to test hopsage seed viability under various storage conditions for a twenty-year period. When the seeds were stored at 15 degrees C. in a sealed container with an agent to keep them dry, germination was better after forty-two months than at thirteen months, increasing from 44 to 57 percent. Even when the seeds were stored at room temperature, germination did not decline but remained at 55 percent. This, of course, is pretty much what would be expected from plants adapted to the dry and uncertain climatic conditions of the desert. At the other extreme, many tropical rainforest plants have a seed viability duration of only a few months or weeks.

Green Molly
Kochia americana

To be strictly accurate, red sage or green, gray, or desert molly, as *K. americana* is variously known, belongs to that group considered to be subshrubs. These are plants with woody bases and with tops which die back during the winter. If we were to include all the subshrubs which occur in the Great Basin, we would have at least four times the number of species in this book. As was pointed out in the introduction, there is not a good dividing line between so-called herbaceous plants and those which are obviously shrubby. Green molly becomes large enough during the summer that it is likely to be considered a shrub by the novice. A common inhabitant of our shadscale desert communities, particularly in dry alkaline flats and saline areas, green molly extends eastward to Wyoming, Colorado, and New Mexico. To the west, however, it is rare outside of the Great Basin. Its common associates, aside from shadscale, include four-winged saltbush, greasewood, spiny hopsage, bud sagebrush, and horsebrush.

Green molly, which gets to be about 30 centimeters high, consists typically of a number of erect, green, frequently pubescent stems growing from a woody base. The leaves are dark green, soft, narrow, and succulent. They vary from 8 to 25 millimeters in length and, when young, may be somewhat pubescent. In some plants the mature leaves retain this pubescent feature.

Kochia is a member of the goosefoot family, the Chenopodiaceae, which along with the aster family dominates the flora of the Great Basin in terms of actual individuals as well as numbers of species. The goosefoot family, a large one with around one hundred genera and over fifteen hundred species, is found on all the continents except Antarctica. Most of the species occur in arid regions, to the extent that in many locales they are the dominant

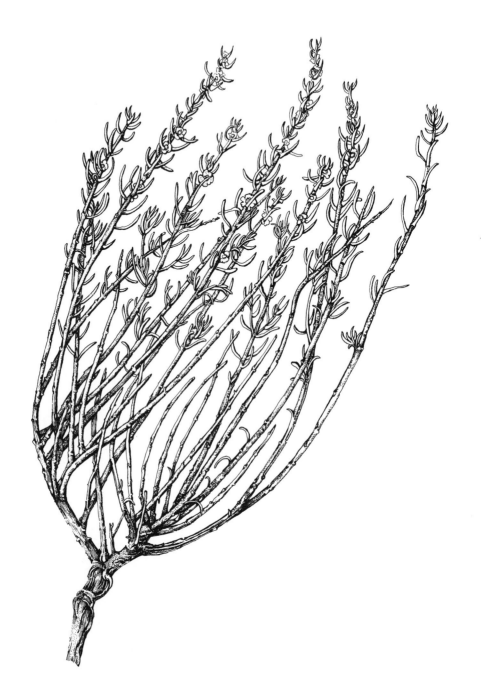

Green Molly

elements of the flora. There are several reasons for their success, and these are discussed in detail in our treatment of shadscale.

Some of the goosefoot genera are trees, many are shrubs, and many are herbaceous plants. The tree and shrub forms are most certainly thought to have evolved from herbaceous forms. As we pointed out in the introduction, the earliest flowering plants were woody, and the herbaceous habit came about by the evolutionary loss of cambium. In the goosefoot family this trend has been reversed, but the wood produced originates in a very different way. In most woody plants that are able to increase in stem diameter, there is only one cambium layer. In the goosefoots, however, concentric cambium layers continue to form, with the result that alternating layers of woody tissue (xylem) and food-conducting tissue (phloem) are formed. Although this may be only an anatomical peculiarity of the family, it may also be somewhat adaptive in that such stems cannot be easily killed by ringing, or removal of the bark in a ring, as is true of most other woody plants. The common garden beet belongs to the goosefoot family, and the concentric rings apparent in a cross section of this vegetable root are also the result of these multiple cambium layers.

The flowers of the green molly are tiny, inconspicuous structures only 1 to 2 millimeters in diameter, located in the axils of the leaves on the upper portions of the branches. The calyx is composed of five sepals fused together at the base, and there are no petals. There are five stamens and a single pistil enclosing only one seed at maturity. Each seed measures about 1 to 2 millimeters in diameter; they are shed from the plant enclosed by the dry, papery pistil and the calyx, which develops wings 2 to 4 millimeters long at maturity.

Green molly is considered fair forage only for sheep and goats, and even for deer it is poor to useless. The genus name honors W. D. J. Koch, a nineteenth-century German botanist. There are about forty species of *Kochia*, primarily in the Old World and Australia. Many are herbaceous and some are annuals—for example, the fire-bush or summer cypress, *K. scoparia*, introduced from Europe and now escaped over much of the United States. An Asiatic species of some forage value, prostrate molly, *K. prostrata*, has been proposed for widespread introduction and may possibly be adventive in Nevada in a few places, though there is no good evidence that it has become established.

Greasewood
Sarcobatus vermiculatus and *S. baileyi*

Near the station is excellent feed for stock, and we stopped here to noon and let our animals feast on the luxurious bunch-grass. Here we were first forced to gather greasewood for our fire, and Tom, being misled by the name, perhaps, traveled off nearly a mile across the plain, momentarily expecting to come upon a grove of large tallowy trees. When he discovered the fact of greasewood being thorny shrubs, the trunks of which were about the size of his little finger, he declared "down upon the infernal prickly yarb."—Dan De Quille, WASHOE RAMBLES

WHILE MANY A modern peregrinator in Nevada would consider Tom's comments not nearly opprobrious enough toward the "infernal prickly yarb," the truth is that without it our alkaline flats would be desolate, lifeless places indeed. Greasewood is able to grow in dense alkaline or saline soils that support little else in the way of shrubs. And, unlike many of our gray-green desert shrubs, greasewood is a bright, almost luminous green of such a characteristic hue that it can be easily recognized, even by travelers in cars whizzing along at freeway speeds.

Greasewood, otherwise known as big greasewood and black greasewood, is an intricately branched shrub from 1 to 2 meters tall, with the ends of the smaller branchlets tapering into sharp thorns. The younger branches are light yellowish or whitish and bear the simple, evergreen, narrow, and fleshy leaves, which are 1 to 3 centimeters long. Although very immature leaves may have some pubescence, this is usually completely lost at maturity. The flowers are very tiny, with the male flowers being borne in small, conelike structures at the ends of the smaller branches. The female flowers are borne singly at the juncture of stem and leaf on the smaller branches back from the

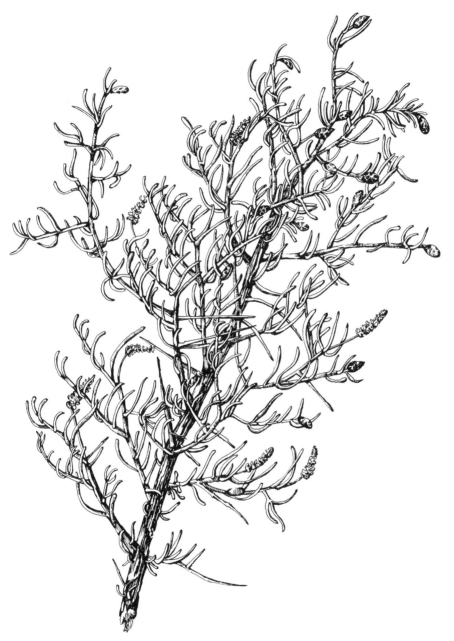

Greasewood, male cones

tip. Individual plants may bear only female or male flowers or both together on the same plant. Each staminate flower consists of just two to five stamens under each scale of the cone. There are no sepals or petals. The female flowers, on the other hand, have sepals fused into a saclike structure which surrounds the single pistil. After pollination, and as the fruits develop, the sepals expand to form a circular wing about 8 to 10 millimeters across, with a point in the center. No other desert shrub in the Great Basin has a fruit shaped in this fashion, although it is similar to that of the Russian thistle, which is, of course, an annual.

Reed W. Fautin, who conducted an extensive study of the northern desert shrub communities in western Utah, considered in detail both the plant and the animal inhabitants of these biotic communities. In one greasewood community, Fautin found that its companions were desert blite, shadscale, green molly, saltgrass, and bud sagebrush. Greasewood, although usually found on alkaline or saline flats, will grow in sandy soils or less saline soils, even occasionally in sagebrush communities, provided sufficient groundwater is available. Another worker, Walter White, found that in the Escalante Valley of Utah the largest communities of greasewood coincided with areas where the water table was less than 15 feet below the surface. He estimated that the shrub could survive even if the water table was as far down as 25 feet. In an investigation of the relationship between various plants and groundwater, Oscar Meinzer discovered that the taproots of greasewood could penetrate from 20 to 57 feet below the surface! It appears that, while greasewood is not an invariable indicator of salinity, its distribution is well correlated with the distribution of groundwater. In fact, when the salinity is above 1 percent, greasewood declines and is replaced by saltgrass, iodine bush, and pickleweed.

James Young and his coworkers recently carried out a study in central Nevada of soil moisture changes in greasewood communities containing islands of shadscale. The penetration of moisture from surface wetting never reached the water table, and greasewood roots were stopped from reaching the water table by permanently dry soil.

Bruce Roundy, James Young, and Raymond Evans studied the development or phenology of salt rabbitbrush, *Chrysothamnus nauseosus* subspecies *consimilis*, and greasewood. Salt rabbitbrush has characteristic green or greenish yellow branches and an elongated, narrow inflorescence, unlike the common subspecies *hololeucus* found in sagebrush associations. Greasewood and salt rabbitbrush are common associates on many alkaline flats of the Great Basin. Roundy and his coworkers found that the phenology of both greasewood and salt rabbitbrush was similar to that of the green rabbitbrush, which had previously been studied. That is, there was first a period of bud burst, generally in late March or early April, followed by a period of restricted growth until mid to late May. Then accelerated growth started. Interestingly enough, this phase ended abruptly during the third week in June for greasewood but continued until early to mid August for salt rabbitbrush.

Roundy's group found that staminate flowers in greasewood began to form in mid May and matured to release pollen in early June. Pistillate flowers on the same plants were not apparent until the staminate spikes had begun to dry, within a few weeks after they had opened. This is a pattern found in most plants which produce both stamens and pistils on the same plant, whether in the same or separate flowers. Frequently it is the pistils which mature first, rather than the stamens. In any event, this mechanism insures that the pistils will not be fertilized by pollen from the same plant and that cross-pollination will occur. Cross-pollination means more variation among the offspring and consequently a greater evolutionary advantage, compared to those forms that are self-pollinated. The botanical term for the maturation of stamens and pistils at different times is dichogamy. If they mature at exactly the same time, the condition is known as synangy.

Joseph H. Robertson, a range ecologist, has worked for many years on the various plant associations of the Great Basin. Recently, he published a comprehensive summary of his own and others' research on greasewood. In reviewing the evolutionary history, Robertson points out that greasewood has been around for a long time—pollen grains from the Eocene epoch, some 50 million years ago, are known. At the present time, greasewood covers some 4.8 million hectares in western North America, more than any other plant whose roots penetrate to the water table (a so-called phreatophyte). Although greasewood inhabits hot desert flats and valley bottoms, Robertson reports that its seeds germinate well only at a relatively low temperature of 11 degrees C. This is an obvious adaptation which insures that seedlings will start growing in the spring rather than the summer, when higher temperatures and the associated higher evaporation rates would be fatal to them. Compared to other desert shrubs, seed production in greasewood is meager, and much reproduction is the result of adventitious buds on the roots, particularly where they have been exposed or injured.

Greasewood, in common with many other alkali-inhabiting shrubs, has the ability to alter the surface layers of soil in its immediate vicinity, probably the result of the accumulation of salt in the leaves and its deposition on the soil surface when the leaves eventually die and fall. Measurements have shown that sodium concentrations become higher under greasewood. T. W. Robinson found that greasewood in Idaho accumulated higher concentrations of sodium, potassium, chloride, sulfate, calcium, magnesium, and boron than the soil in which it was growing.

A very closely related species is Bailey's greasewood or little greasewood, *S. baileyi*. It differs primarily in its relatively small stature, about 30 to 60 centimeters; its branchlets and leaves are minutely grayish pubescent; and its fruits have a wider wing. Bailey's greasewood grows in much drier sites and is a common associate of shadscale throughout the Great Basin. It is easily distinguished, even at a distance, from big greasewood by its somber gray aspect, and it is not considered to be a phreatophyte. The only other desert shrub with which it might be confused is Shockley's desert thorn, *Lycium shockleyi*, which belongs to an entirely unrelated family. Shockley's desert thorn produces berries, not the winged fruits of Bailey's greasewood, and the former has wider, spatulate leaves, with the ultimate branchlets somewhat stouter and more angular. Desert thorn becomes larger and tends to grow on better-drained slopes than Bailey's greasewood, although the two can frequently be found as associates in transitional areas. There are many botanists who regard Bailey's greasewood as only a minor variant, not deserving of species recognition, since many intergrades can be found between the two greasewoods. As we will see in the case of other shrubs, such as the sagebrushes, the concept of species is a human invention, and every botanist has a personal interpretation of the concept.

As might be expected, the range manager regards greasewood as being of little or no value to most domestic livestock. Sheep and goats make some use of it, and it is considered good forage for these animals. It is a dangerous browse under some conditions, however, since oxalates are known to accumulate in the leaves—horses, cattle, and sheep have died from eating too much of the foliage. As is true of our other shrubs, greasewood forms an integral part of a plant community which supports a variety of wildlife, in-

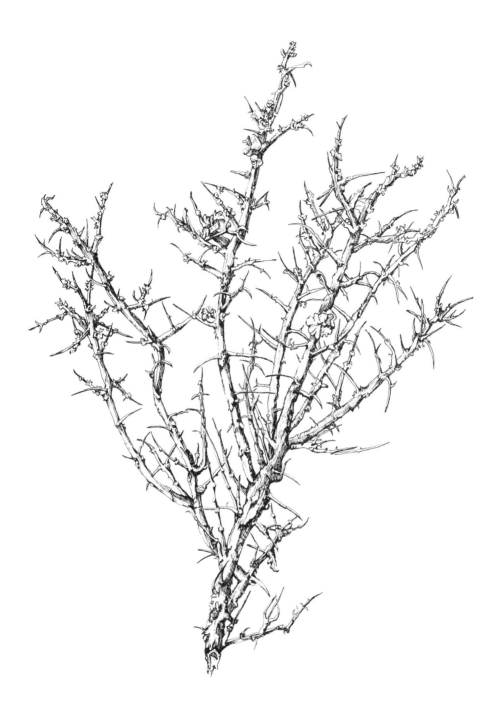

Bailey's Greasewood

cluding our ubiquitous jackrabbit as well as other mammals common to the shadscale deserts and alkali flats. Indirectly, of course, it contributes, through the processes of growth and decay, to myriad forms of animal life from insects to lizards and birds.

The genus *Sarcobatus* contains only the two species, big greasewood and Bailey's greasewood; it ranges from Mexico north to Alberta, Canada, and east through the Great Basin to South Dakota, Colorado, and western Texas. The range of Bailey's greasewood, somewhat less than that of the big greasewood, extends from California through Nevada to Colorado.

The genus name is derived from the Greek *sarx*, meaning flesh, and *batos*, meaning bramble. The species name *vermiculatus* comes from the Latin *vermis*, meaning worm, and *culus* or small. The term *baileyi* was conferred in 1892 in honor of Liberty Hyde Bailey, the famous horticulturist and botanist of Cornell University.

Desert Blite
Suaeda torreyana

THE DESERT PLAYAS of the Great Basin are in many ways as hostile to higher plant life as the bleakest Antarctic landscape. Frequently there grows a thin stand of scraggly shrubs at the edge of and partly onto the larger playas. On examination, this almost always turns out to be the desert blite, *S. torreyana*. Neither heat nor lack of water is the barrier to colonization in such areas by higher plants. It is, rather, the extremely high salt content of the soil that, in many instances, totally inhibits the growth of all higher plants. Some halophytic bacteria and algae have solved this physiological problem, but mosses, ferns, and all higher plants are totally absent from the most saline locations.

Because of the tremendous plasticity of living forms on this earth, it is tempting to believe that, provided it is not too hot or too cold, living things can eventually adapt to any habitat. This is a generalization which, if restricted to the most primitive forms, holds well enough. But apparently there has not been enough time since seed plants first evolved, some 300 million years ago, for them to solve this particular problem. Some flowering plants have gotten back into salt water, notably such forms as the eel grass and turtle grass that thrive on the continental shelf. But they have very few competitors among higher plants—only several among the hundreds of higher plant families have been able to make the transition to salt water. In the case of desert playas, the salt concentration is frequently much higher than that of the oceans, and the situation is additionally complicated by the lack of water at some time during the year.

We now think that some of the difficulty lies with the poisonous nature of sodium in such soils, at least for certain plants. Plants such as the shadscale, tamarisk, and eel grass that can resist relatively high salt concentrations are

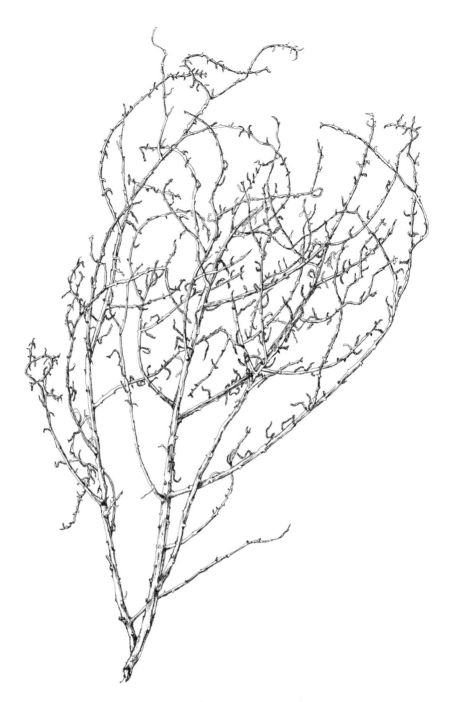

Desert Blite, winter aspect

apparently able to do so as a result of a cellular mechanism that accumulates and excretes salt. This excretion requires a good deal of cellular energy as well as a complicated structure, which most plant families have not evolved. However, some salt-tolerant plants, such as the desert blite, are not able to excrete salt but simply continue to accumulate it in their leaves during the growing season. When the leaves die and are shed at the end of the season, the accumulated salt is also shed. Year after year of this sort of thing results in the surface layers of the soil having a much higher salt concentration, to the extent that few other plants are able to get started as seedlings.

The ferns, which have been here at least 100 million years longer than the seed plants, have never managed to evolve a salt tolerance. And there are no saltwater mosses. Since the earth is destined, according to some, to become drier and more saline, our desert shrubs, with their adaptations to this extreme environment, may give us a glimpse into the future possibilities for evolution in higher plants. The goosefoot family, to which the desert blite belongs, is certainly the most successful land plant family in solving the twin problems of aridity and salinity; it may well be the model for future flowering plants on the earth!

Desert blite varies in height from 30 centimeters to 1 meter. Its branches are only sparsely leafy, at best, with flat, narrow leaves about 2 millimeters wide and 30 millimeters long. The flowers are very small, greenish, and inconspicuous, being only 2 to 3 millimeters broad. The five-lobed calyx encloses the fruit at maturity. There are no petals, five stamens (sometimes fewer), and one pistil containing a single seed. If the flowers are closely examined, some will be seen to lack stamens while others lack pistils, even though most of the flowers will have both. A variable pattern such as this can be regarded as the first step on the road to a species with the sexes completely separated on different plants, as is true of many other members of the goosefoot family.

Desert blite, as well as other related species, was utilized by the Indians to prepare a tea which they believed was useful for kidney problems. Additionally, it was thought that the juice from the plant helped alleviate itching.

The genus name, *Suaeda*, is an old Arabic name for the North African species. There are about fifty species, found worldwide, in the genus. The species is named after John Torrey, a famous American botanist of the nineteenth century who worked on many of the plants brought back by John Frémont and other explorers of the West. At Columbia College in New York, Torrey founded what became the Torrey Botanical Club in the year of his death, 1873.

There is a variety of *torreyana* called *ramosissima*, distinguished by finely pubescent young stems and leaves. It is found throughout much of the range of the typical form which extends from eastern Oregon to Wyoming and south to Texas and California. Another species found in the Great Basin is S. *fruticosa*, the alkali seepweed, characterized by completely hairless stems and leaves which have a waxy, bluish bloom. The leaves, in addition, are cylindrical and succulent rather than flat, and those located among the flowers on the ends of the branches are much smaller than those on non-flowering branches. Alkali seepweed extends north to Alberta and south into Mexico. This is also an Old World species, found throughout the desert areas of Eurasia and Africa. There are several other species of *Suaeda* in the Great Basin, but they are annuals.

POLYGONACEAE
BUCKWHEAT FAMILY

Kearney's Buckwheat
Eriogonum kearneyi

ALMOST AS characteristic of the Great Basin as shadscale and sagebrush are the various species of wild buckwheat. Some of the latter are annual or perennial herbaceous plants, but many are woody, at least at the base. Some authorities call them subshrubs. If we were to discuss all these partially woody forms, we would have to consider perhaps thirty-five species within the Great Basin. Fortunately, most of these are low, compact forms that, to the uninitiated, appear to be wildflowers rather than shrubs. Separating many of these into species is a job best left to the expert. Consequently, in this book we shall confine our discussions to those wild buckwheats which are pretty obviously shrubs, and save the others for the wildflower book in this series.

Kearney's buckwheat is one of the tallest in the genus. It frequently attains a height of over 1 meter. Outside of the Basin in southern Nevada, the California buckwheat, *E. fasciculatum*, not uncommonly gets to be 2 meters and sometimes 3 meters tall. Confined primarily to sandy areas, Kearney's buckwheat is a conspicuously large and diffusely branched shrub. In areas with sand dunes, clumps may be several meters in diameter on the tops of small dunes. Obviously, they assist to some extent in stabilizing dune areas, though they are not as important as certain other dune plants.

The pubescent, 1-to-3-centimeter leaves are sparsely scattered along the lower stems, which are also pubescent. The upper part of the stem tapers to a many-branched inflorescence which bears numerous small, whitish flowers. Examination of these flowers under a hand lens will reveal the pattern characteristic of all the buckwheats: two whorls, consisting of three whitish sepals each, enclosing nine stamens (sometimes fewer) and a single

Kearney's Buckwheat

pistil in the center. There are no petals and, even though the sepals are colored like petals in the buckwheats, they are structurally not petals. Long ago in the evolution of the buckwheats, the petals, for some reason, were lost. And, as we have seen elsewhere, "nature" changed its mind, decided that petals were needed once again, and formed them from sepals. Of course, nature doesn't have a mind to change; this is but one more example of the randomness of evolution. It also illustrates the fact that, once the genes for a particular structure have been lost, they can't be resurrected. The best that can be done is to take the genes for another structure, in this case the sepals, and change them to produce something that resembles the lost petals. But, of course, this is all the result of natural selection and the appropriate hereditary variations—it is not a purposeful evolution of these structures on the part of the plant.

The pistil in the wild buckwheats matures into a triangular fruit called an achene. Except for its much smaller size, its appearance is very much like that of the cultivated buckwheat achene.

The family to which the wild and cultivated buckwheats belong is the Polygonaceae. There are around thirty genera and over a thousand species in this largely temperate-zone family. Incidentally, although most members of the family have the flower parts in threes, as do the buckwheats, the family does not belong to that major group of flowering plants called the monocots. Virtually all beginning botany texts recite the litany that flower parts are in threes in the monocots and in fives in the dicots. But, as with most such generalizations, there are always exceptions. In fact, a few members of the buckwheat family do have five petallike sepals (but no true petals). Interestingly enough, these five-part flowers have evolved from the usual two whorls of three sepals by a fusion of one member of the inner whorl with one member of the outer whorl. Threes and fives seem to be magic numbers in the flowering plants!

The genus name *Eriogonum* comes from two Greek words, *erion* which means wool and *gonu* which means knee. Many of the buckwheats have a woolly pubescence, and, like other members of the same family, one of their distinctive features is a stem node which is characteristically swollen and sometimes bent. In some *Eriogonums* these nodes are especially hairy, accounting for the genus name. The genus consists of perhaps 150 species confined to North America, with most of the species being in the West. The species name honors Thomas H. Kearney, a botanist and cotton breeder who, along with Robert Peebles, authored a significant state flora, *Flowering Plants and Ferns of Arizona*, published in 1942.

Great Basin Buckwheat
Eriogonum microthecum

THE GREAT BASIN buckwheat is the most common shrubby species of this genus found throughout the arid portions of the Great Basin. Sometimes known as the slender buckwheatbrush or Nuttall's buckwheat, it superficially resembles the rock buckwheat. However, the Great Basin buckwheat occupies a considerably wider amplitude of habitats, being found in large numbers not only on dry, rocky slopes but also on deeper, sandy soils. As an abundantly branched shrub, it may attain a height of 30 centimeters, though more frequently it is about 20 centimeters high.

Several important differences help separate the Great Basin buckwheat from the rock buckwheat. Great Basin buckwheat has smaller leaves, generally about 1 to 1.5 centimeters long, elliptical in shape, white-pubescent beneath, and smooth on the upper surface. The most distinguishing feature, however, is the inflorescence, in which the individual flowers are borne in a flat-topped, repeatedly branched structure up to 5 centimeters wide, with the smaller branches borne opposite one another. In the rock buckwheat, the flowers are crowded together in a globose cluster. Also, unlike the latter species, in the Great Basin buckwheat there is no whorl of leaflike bracts about the middle of the flowering stems. The dead flowering branches frequently persist until the next spring, allowing for easy identification even in the middle of winter.

Individual flowers, like those of the other buckwheats, consist of six small, white or yellow sepals about 2 to 3 millimeters long, no petals, nine stamens, and one pistil. Several flowers are borne together in a cuplike, lobed structure (called an involucre) about 3 millimeters long. Such involucres are an important diagnostic feature of the buckwheats.

Great Basin Buckwheat

Heermann's Buckwheat

The Great Basin buckwheat's range extends from southern California to eastern Washington and east to Colorado and New Mexico.

Another shrubby buckwheat common as an occasional plant here and there in the shadscale desert of the western Great Basin is Heermann's buckwheat, *E. heermannii*. It is easy to recognize because of the persistent flowering stems, which branch in a very regular, intricate fashion. Each flowering stem at its base divides into two equal branches, each of these in turn divides into two, and so on. Some very profusely branched specimens resemble nothing so much as a giant reindeer lichen. This kind of regular branching, known as dichotomous, is rare among flowering plants but common in many nonflowering, primitive groups. However, this does not imply that this branching pattern should be regarded as a primitive feature which has been retained by some of the buckwheats. As we have seen in other cases, certain apparently primitive features are frequently just the result of evolution reintroducing a feature which appeared to have been lost. And, usually, anatomical traces of the more "advanced" ancestors can be found if the plant is studied under the microscope, thus distinguishing it from truly primitive forms.

Heermann's buckwheat has smooth upper branches, while the lower ones have patches of woolly pubescence. The leaves are white-pubescent underneath and resemble those of the Great Basin buckwheat. The individual flowers are pale yellow; several are borne together in involucres at the ends and on the upper forks of the flowering branches. Heermann's buckwheat frequently inhabits drier sites than either the rock or the Great Basin species; it ranges from southern California and southern Nevada north to Humboldt County, Nevada.

Great Basin buckwheat is considered to be of some value to cattle and sheep. During the winter, it is regarded as a fair to good food plant.

The Great Basin buckwheat's specific name means small box, because of a fancied resemblance to boxwood. Sometimes it is called Nuttall's buckwheat, in honor of the British-born botanist Thomas Nuttall, who explored much of the Americas for plants during the early part of the nineteenth century. The original specimen of Great Basin buckwheat from which he drew up his description came from near Walla Walla, Washington. From 1822 to 1834, Nuttall served as curator of the Harvard Botanical Gardens.

Rock Buckwheat
Eriogonum sphaerocephalum

AS THE NAME implies, dry, rocky slopes and ridges are the habitat of the rock buckwheat. Another name for it, this one derived from the Latin species name, is round-headed buckwheat. Rock buckwheat is a low shrub, rarely over 10 centimeters high, found abundantly in the sagebrush and pinyon-juniper communities throughout the Great Basin.

The 1-to-3-centimeter, somewhat spatulate leaves tend to be whorled near the ends of the branches. Generally the leaves are densely white-pubescent beneath, while the upper side varies from smooth to very pubescent. The flowering stems, which are typically 5 to 10 centimeters long, have a whorl of leaflike bracts a little above the middle. A cluster of fused bracts with seven or eight lobes immediately supports the spherical cluster of creamy to yellow flowers, produced from May to July, depending on the elevation. This cluster of fused bracts, the involucre, is especially characteristic of many of the wild buckwheats. Individual flowers are composed only of sepals, as described for Kearney's buckwheat.

Rock buckwheat ranges from western California, north to Washington, and east through the Nevada and Idaho portions of the Great Basin.

Another very distinctive and easily recognized low, shrubby wild buckwheat occurs on rocky to fine gravelly slopes in the extreme western Great Basin. Wright's buckwheat, *E. wrightii* variety *subscaposum*, typically hugs the ground, forming a dense mat of short branches covered with small, eliptical leaves varying in length from half a centimeter to a centimeter. Sometimes older plants will form conspicuous, dense mounds up to half a meter in diameter. Wright's buckwheat, fortunately, is about the only wild buckwheat in the Basin to attract any significant commercial interest. Plants are dug up, dried, and either left in their naturally gray aspect or dyed some

Rock Buckwheat

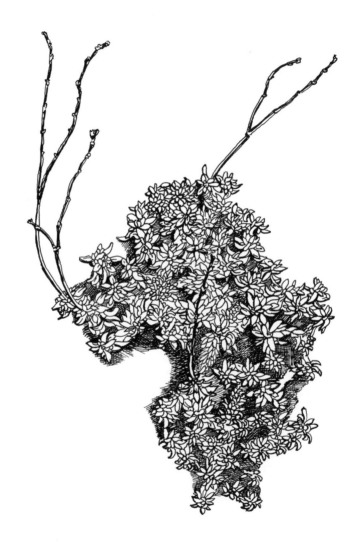

Wright's Buckwheat

Altered Andesite Buckwheat

bizarre color and mounted in suitable containers to be sold as "Ming" trees. The woody rootstock or caudex, when exposed, frequently resembles the twisted and gnarled trunk of a very old bonsai. Although Wright's buckwheat is not rare or even uncommon, too much commercial interest of this sort could conceivably reduce its numbers significantly, since only wild plants and not cultivated ones appear to be used for this purpose.

In the western Great Basin, Wright's buckwheat is particularly common on areas of so-called altered andesite in the Carson Range between Reno and Virginia City. These altered andesite areas have resulted from the changes produced in igneous deposits subjected to subterranean water under considerable heat and pressure. Once exposed and weathered, such rocks tend to form soils poor in nutrients and strongly acid. Altered andesite areas are easily recognized, even from a distance, since their soils favor the growth of ponderosa pine rather than the pinyon-juniper of surrounding communities. The absence of sagebrush and many other shrubs from altered andesite regions causes them to appear particularly barren.

The several varieties of Wright's buckwheat occur in the various mountain ranges throughout California; the variety *subscaposum* is found from 5,000 to 11,000 feet in the Sierra Nevada south to the mountains of southern California.

Another wild buckwheat found only on these acid soils of the extreme western Great Basin is the altered andesite buckwheat, *E. lobbii* variety *robustum*. Found only in Storey and Washoe counties in western Nevada, the altered andesite buckwheat was, a few years ago, regarded as threatened. However, the consensus now is that its unique habitat does not appear to be in imminent danger from developers. This buckwheat is so distinctive that it is not easily confused with any other species in the same area. Its stout woody rootstock bears round or oval leaves with long petioles and blades between 1 and 4 centimeters long; the leaves are borne in a tuft at ground level and are very densely pubescent, appearing a light gray-green color. The leafless flower stalks bear clusters of cream-colored flowers held well above the leaves.

There are many other subshrubby species of buckwheat to be found in the Great Basin, but to describe them all is beyond the scope of this work, and to identify them usually requires an expert on the group.

TAMARICACEAE
TAMARISK FAMILY

Tamarisk
Tamarix spp.

The Tamarisks are all of graceful and distinctive appearance, with light and feathery foliage and large, loose panicles of pinkish flowers. . . . As they are inhabitants of warmer arid regions, they are well adapted for countries of similar climatic conditions. They are also excellent for seaside planting. They grow well in saline and alkaline soil and thrive in the very spray of the salt water. —Liberty Hyde Bailey, CYCLOPEDIA OF AMERICAN HORTICULTURE

THE GENUS *Tamarix* occurs naturally from western Europe and the Mediterranean to North Africa, northeastern China, India, and Japan. The number of species is variously estimated to be between fifty and ninety. Bernard R. Baum of the Plant Research Institute in Ottawa, Canada, wrote a monograph on the group a few years ago that recognized fifty-four species. Of these, eight have become naturalized in North America, primarily in the Southwest, though one species is grown as far north as southern Canada.

The colonization of the Great Basin by tamarisk is dramatically portrayed in a research paper by Earl Christensen entitled "The Rate of Naturalization of *Tamarix* in Utah." Christensen looked at the historical records for Utah Lake, the Great Salt Lake, the Colorado River, and the Green River. He could find no evidence that tamarisks were present at any of these locales prior to 1925. As late as 1934, Seville Flowers, in his publication on the vegetation of the Great Salt Lake region, makes no mention of its occurrence there. But twenty years later it was common. Similarly, tamarisks were recorded at Utah Lake in 1926 and became common in the period from 1930 to 1942. Now, throughout the Great Basin, they are abundant in wet

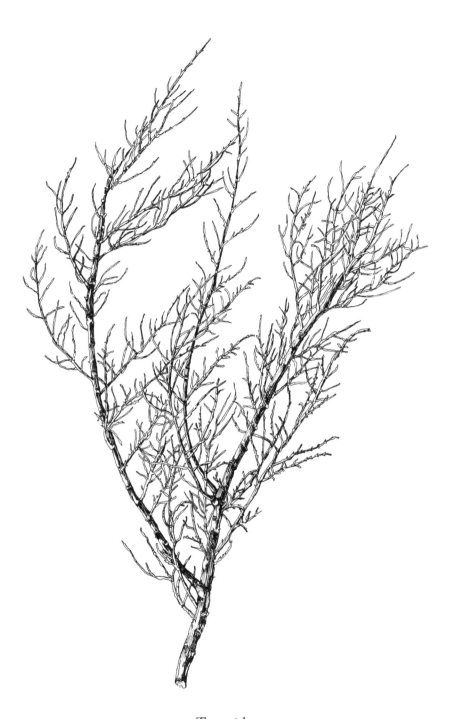

Tamarisk

areas, whether saline or not. Outside of the Basin, Ivar Tidestrom recorded tamarisk along the Virgin River near Saint Thomas, Nevada, in 1919.

Tamarisks are easily recognized by their small, scalelike leaves, very much like those of juniper, though smaller, and their slender, supple branches. They are, in fact, known as salt cedars by many. Linnaeus, the father of the so-called binomial system of naming plants, mistakenly described one tamarisk as a species of arborvitae. Several species are so well preadapted to Great Basin saline and alkaline soils that to the uninitiated they might well appear to be native shrubs. A traveler along the western shore of Walker Lake will see concentric rows of virtually pure stands of tamarisk, each row marking a former shoreline of this rapidly disappearing lake. The species along Walker Lake is known as *T. parviflora*. This is the easiest species to recognize when in bloom, since its tiny, pink flowers have only four petals. All other introduced tamarisks in the Basin have five petals. In the same general area can be found a species known as *T. chinensis*. Another species, *T. ramosissima*, appears to be a little less common. It can be found in the Fernley Wildlife Management Area and around the Soda Lakes in Churchill County. These last two species are rather difficult for the novice to separate. *T. parviflora* is found in southern Europe and Asia Minor from Yugoslavia to Turkey. *T. chinensis*, as might be guessed, occurs from Mongolia and China to Japan. *T. ramosissima* is found from the Ukraine and Iraq east through China and Tibet to Korea.

Tamarisks were originally planted as ornamentals and as windbreaks, but their profligate nature soon demonstrated that all such introductions should be carefully monitored if major ecological catastrophes are to be avoided—witness the Australian gum trees and Scotch broom in California or halogeton and Russian thistle in the Great Basin. Some introductions, of course, are accidental, as was the case with Russian thistle, and probably such errors are inevitable given the nature of modern commerce. However, some problems that have resulted from deliberately introduced forms might have been avoided if sufficient research had been done beforehand. In many instances, introduced species have grown out of bounds because the natural enemies which kept them in some sort of balance in their homelands are missing. The result is that many native species have been crowded out of some areas. Angus Woodbury, Stephen Durrant, and Seville Flowers found that, in 1958, tamarisks made up 19 percent of the total vegetation cover along the streamside of Glen Canyon. They found that tamarisks usually occupied the sandier areas and willows occupied the muddier sites. They concluded that competition between the two would eventually restrict the willows to the muddy banks, with the sandy areas being occupied exclusively by tamarisks.

Individual flowers in *Tamarix* are quite small, with petals from 1 to 3 millimeters long. There are five stamens, except in *T. parviflora*, which usually has only four. Individual flowers are borne on short stalks arranged along the axis of a terminal branchlet. Interestingly enough, there are significant differences between flowers produced in the spring and those formed during the summer. In some species, the inflorescences formed in the spring are denser than those developed during the summer. There are also other differences, mainly in the arrangement of the stamens and the size of the petals, which tend to be bigger in the summer flowers. The flowers are bisexual, and subsequent to pollination small capsules are produced with seeds bearing a tuft of hairs.

Those plants, such as tamarisk, which are able to grow in saline conditions are termed halophytes. A variety of adaptations have enabled certain higher plants to accommodate to high concentrations of salt in the soil. In the case of tamarisk, salt is accumulated by specialized glands sunk in depressions on the leaf surface. This salt is then excreted onto the leaf surface. Because of this ability, tamarisk is known as a salt accumulator. The accumulation and excretion of salt take a great deal of energy—consequently, it is not surprising that the cells of salt glands are packed with those small structures called mitochondria which are universally responsible for respiration, that is, turning food energy into other usable forms of energy. The ratio of the various salts excreted is, oddly enough, not directly dependent on the ratio that may occur in the soil. Tamarisk secretes sodium preferentially, even though it may be no more concentrated than other elements, such as potassium, in the soil.

At one time it was believed that most plants were not able to grow under saline conditions, primarily because the total concentration of salts per unit volume was greater outside the plants than within them. Presumably, this meant that plants would lose water to the soil by a process known as osmosis, since the net movement of water is always from an area of lower salt concentration to a region of greater salt concentration. It is this phenomenon that allows us to preserve pickles, fish, or what-have-you in brine solutions. Decay bacteria cannot grow because they lose water to their environment. However, a primary problem with higher plants in moderately saline habitats now appears to be that, above relatively low levels, sodium is poisonous to plants. In fact, for most plants, sodium is not essential to growth. The exceptions would be some species of blue-green algae and at least some forms of saltbush.

Tamarisks may be further adapted to dry, desert conditions by their ability to absorb water from the air when the humidity is high, though possibly this is of value only when the plant is somewhat wilted. Another adaptation, shown by this and other genera of desert shrubs and trees, is their ability to make use of fog. Fog droplets collide with branchlets, accumulate, and eventually fall as drops to the ground. Y. Waisel discovered that, in the Negev Desert of Israel, salt-secreting *Tamarix* was better able to trap fog than other tree species in the area. He found that the soil beneath such trees had been moistened to a depth of 50 centimeters. Another aspect of this secretion is that salt tends to accumulate in the surface layer of soil. This, of course, makes the salt-accumulating halophytes even better able to compete with other plants.

Tamarisks are of no direct value to livestock, and one might suppose that they are generally useless, except as ornamentals or for erosion control. One species, however, the Athel tamarisk, *T. aphylla*, introduced into the warmer areas of the Southwest from North Africa, gets to be a fair-sized tree 18 meters tall. Its wood is fine-grained and light-colored, and G. E. P. Smith of the University of Arizona proposed some years ago that it would be useful for cabinet wood and fence posts. According to Lyman Benson and Robert Darrow in their recent book, *The Trees and Shrubs of the Southwestern Deserts*, most of the examples of the Athel tamarisk in the Southwest have been propagated from half a dozen Algerian cuttings shipped to a botanist, J. J. Thornber, at the University of Arizona about eighty years ago.

Another species, *T. mannifera*, found throughout the Middle East, is frequently infested with a scale insect which causes the production of a sweet, white, gummy material called manna. This may be one source of the manna referred to in the Bible. Insect galls, which occasionally develop on the tamarisk, have been used as a source of medicines and tannin in the Old World.

Tamarisks belong to the family Tamaricaceae, consisting of four genera and about one hundred species generally found within drier, saline habitats in the subtropical and temperate zones of Africa, Europe, and Asia. However, tamarisk also grows in maritime locales as far north as Norway. No members of the family are native to the New World. The species name *parviflora* means small flower, while *ramosissima* means with many branches.

SALICACEAE
WILLOW FAMILY

Coyote Willow
Salix exigua

Of the Willow or Sallow, eight kinds thereof. . . . surely there is not more profit arising from any other tree of the waters than from it. . . . Ye shall have of these osiers, some that are very fine and passing slender, whereof are wrought pretie baskets and many other daintie devises; others that are more tough and strong, good to make paniers, hampers, and a thousand other necessary implements for countrey houses. —Pliny, HISTORIA NATURALIS

THE GREAT BASIN INDIANS as well as the Romans were familiar with the manifold uses of willow. Its supple branches, when green, could readily be woven into baskets and other useful items. Of all our willows, the coyote willow is the best for such purposes, for its slender, straight, reddish branches—up to several meters in length—are ideally suited to the artisan.

It is improbable that an irrigation ditch or a stream exists anywhere in the Great Basin without some coyote willow along it. The coyote is the only willow that can be found in the lowest and hottest portions of our deserts, provided its roots are in moist soil. Its adaptation to a drier environment is apparent in its narrow, pointed, and gray-pubescent leaves, which are 5 to 14 centimeters long but only 3 to 10 millimeters wide. No other willow in the Great Basin has leaves with this combination of characteristics. In very moist habitats in the mountains, the leaves will not be as pubescent, and they will be somewhat wider. Whether this is the result of a genetic or an environmental difference has never been investigated.

As with all willows, the sexes of the coyote willow occur on separate plants. Catkins containing the very much reduced flowers are developed after the leaves have begun to unfold in the spring. The male flowers consist of one bract and two stamens, while the female counterpart is comprised of a bract and a single pistil. There are no sepals or petals, and at the base of each bract are one or two small glands. There is some indication that these glands are all that is left of the calyx. The female catkins are relatively inconspicuous, while the male catkins, because of the silky hairs associated with them, are prominent.

If one considers only the structure and the inconspicuous nature of the willow flower, a logical assumption would be that, like many other catkin-bearing plants, it is wind-pollinated. However, this is not the case with most willows, which are insect-pollinated and commonly visited by bees. Undoubtedly, this is a so-called secondary characteristic for willows. That is, their ancestors at one time were wind-pollinated, but they evolved once again in the direction of insect pollination. (We think that the earliest flowering plants were insect-pollinated.) This is but one more example of the random aspect of much of evolution.

After fertilization, a capsule develops which eventually splits open into two valves and releases the numerous tiny seeds, each with long hairs that aid in dissemination by the wind. The short-lived seeds must land in a suitable location and germinate immediately. Coyote willow also extends its territory by means of creeping rootstocks, which produce numerous upright stems to form dense thickets.

As an undergraduate student a very long time ago, this author was taught that the willows were a primitive family because they had such simple flowers. My mentor was still an adherent of the Engler-Prantl system, which says that the fewer parts a flower has the more primitive it is. Adolph Engler, professor of botany at the University of Berlin, and his associate Karl Prantl published a twenty-three-volume work from 1887 to 1915 called *Die Natürlichen Pflanzenfamilien*, which included all the genera of plants known at that time, from algae to the most advanced flowering plant families. This monumental work influenced botanists the world over, and at least for a time they accepted the view that the first flowering plants had very simple flowers with no sepals or petals, few stamens, and few pistils. All the catkin-bearing types such as oaks, walnuts, and willows were placed in a larger group called the Amentiferae.

Without going into the complex studies which took place in developmental plant anatomy, paleobotany, and biochemistry since the turn of the

Coyote Willow, female catkins

century, it became apparent by the early 1930s that Engler and Prantl had in effect gotten part of the evolution of flowering plants backward, for the earliest flowers were, we now believe, much more like magnolia flowers—having numerous poorly differentiated petals and sepals, as well as numerous stamens and pistils—and were pollinated by insects (probably beetles). Evolution then proceeded from this basic type to produce the very complex orchid and aster flower forms, but evolution doesn't always result in more complex forms—sometimes it simplifies. This makes the job of the evolutionary biologist much more difficult: is a particular feature simple because it is primitive, or was it once a complex structure which has become simpler over time? There is no question that, in some lines of seed plants, evolution has tended toward the reduction or elimination of flower parts to produce a simpler design. In the case of the willow, this reduction has obviously gone about as far as it can go! Another indication of the relatively advanced position of the willow is the fact that its wood shows advanced anatomical characteristics.

The idea that the type of flower exemplified by the magnolia is much like the primitive flower type which came into existence sometime over 100 million years ago was originally proposed by Charles Bessey, a professor of botany at the University of Nebraska. His full treatment of this theory was published in 1915 by the Missouri Botanical Garden. Since that time the theory has been somewhat modified, but the basic idea has remained the same, and it has greatly influenced the thinking of botanists the world over. Charles Bessey can rightfully be considered the Einstein of systematic botany, to whom we all owe an intellectual debt.

At the time that the flowering plants were first evolving and competing with the prevailing conifer forests (during that period known as the Cretaceous), the Great Basin had a warm-temperate to subtropical climate, and plants such as the fig, cinnamon tree, magnolia, laurel, and palm flourished in what is now sagebrush and pinyon-juniper habitat. Apparently there was no frost, and the heavy rainfall was distributed more or less evenly throughout the year. Beginning about 70 million years ago, the Great Basin began to get cooler and drier, although this trend was interrupted by many warm periods. In retrospect, the present botanical aspect of the Basin is a very recent one and, from a geological perspective, probably a fleeting one.

While we appear to have gotten somewhat off the track, such is not entirely the case, for there were willows in the Great Basin during the Cretaceous—and, probably, they have always been here since then, even though our tropical friends have left the scene. We have about twenty species of willow here, some of which are small trees. All of them grow in moist areas, and most are in the mountains. Unfortunately, separating them into species is, in general, not easy. In many instances, one must have the catkins to be certain of the identity of a particular example.

Willows produce a compound known as salicin which is chemically closely related to acetylsalicylic acid, otherwise known commonly as aspirin. Salicin may account for the numerous pharmaceutical properties attributed to willows by the Indians of the Great Basin. They used various preparations from willows, probably mostly the coyote willow, to treat toothache, stomachache, diarrhea, dysentery, venereal disease, and even dandruff!

Willows are regarded as important range plants, although palatability varies with location and species. Apparently, cattle find them much more palatable in late summer than earlier in the season. Even the fallen and brown leaves are eaten by sheep. Willow is one of the favorite foods of beavers in the Great Basin, and it appears to be an important browse plant for deer.

The willow family, Salicaceae, has about three hundred species widely distributed throughout the north temperate region and into the Arctic, where forms less than 10 centimeters high exist. Interestingly enough, there are no native willows in Australia. The only other genus in the family is the poplar, *Populus*, with about forty species. All the poplars, unlike the willows, are wind-pollinated. There is good reason to regard the willow family as a member of the same order which includes the violet family, a relationship which is certainly not apparent to the general observer.

The genus name *Salix* is an old Latin name for willow, which is thought to be derived from two Celtic words, *sal*, meaning near, and *lis*, meaning water. The species name *exigua* means small.

BRASSICACEAE
MUSTARD FAMILY

Bush Peppergrass
Lepidium fremontii

ALTHOUGH MANY MEMBERS of the mustard family are denizens of our deserts and mountains, the bush peppergrass is the only significant woody species here. In fact, only a few members of the genus to which it belongs are shrubs, and many are annuals which complete their life cycle within only a few weeks during spring in the desert. Actually, only the basal parts of the bush peppergrass are woody, and by some this plant would be considered to be a subshrub. Other names in common use are desert alyssum and Fremont's peppergrass.

L. fremontii is a low, rounded, many-branched shrub with whitish, waxy stems; it is frequent in sandy and rocky soils in the southern Great Basin, continuing on south into creosote bush and Joshua tree country. Generally, it is found between 2,000 and 5,000 feet. Although most manuals consider this a shrub of the southern deserts, in the western Great Basin it occurs as far north as the Granite Range in northern Washoe County. It is fairly common in shadscale desert areas in several western Nevada counties, especially in the vicinity of Pyramid Lake. The only other peppergrass with which this species might be confused is *L. montanum*, the mountain peppergrass. Both species may grow to a height of 20 or 30 centimeters, although the mountain peppergrass tends to be smaller, has fewer stems, and is less woody than the bush peppergrass. In addition, *L. montanum* is distributed from sagebrush areas out into the shadscale desert, and its range includes all of the northern Great Basin.

With a hand lens, one can always separate these two species with ease, since the variety of the mountain peppergrass found predominantly within the Basin, *canescens*, has finely pubescent stems, while the bush peppergrass is always hairless. As the species name implies, *L. montanum* occurs at somewhat higher elevations, from 4,500 to 6,500 feet, although it is not really a mountain inhabitant.

Bush Peppergrass

The genus name *Lepidium* comes from the Greek *lepidion*, which means little scale. This is in reference to the small, flattened, circular seed pods. The species name *fremontii* honors the explorer John Charles Frémont, who dubbed the area he had just traversed the "Great Basin" in 1844.

The flowers in both species are white and borne on slender stalks along the axis of the inflorescence; the petals are only 2 or 3 millimeters long. Blooming occurs from spring to early summer. Following flowering, small, flattened, circular fruits develop. In the bush peppergrass the fruits have a wide, winged margin and are 5 to 6 millimeters long, while mountain peppergrass fruits have only a narrow margin and are only half as long. The stem leaves in the bush peppergrass are narrow and long, about 2 to 5 centimeters, with the lower ones sometimes possessing narrow lobes attached to the main axis.

The mustard family, to which *Lepidium* belongs, is now called the Brassicaceae, although until recently it was known as the Cruciferae, in reference to the cross formed by the petals. All the mustards are characterized by a very regular arrangement and number of floral parts. They consistently have four sepals, four petals, six stamens, and one pistil. Because there is so little variation from this basic theme, this is one of the easiest families to recognize in the field. Not only the flowers but also the fruits of this family are distinctive, consisting of either elongated capsules, with the halves splitting away from each other to release the seeds, or short and frequently circular capsules like those of peppergrass, which contain fewer seeds but which nevertheless split open to release them in the same fashion. Structurally, the capsules are divided into halves by a membrane running down the center. Often, this central membrane remains attached to its stalk long after the seeds are shed. This is a distinctive feature, common to most mustards but infrequent in other families, at least in our area. Consequently, it is usually possible, even in the dead of winter, to recognize a mustard, though only a few lifeless stalks may be evident.

The flowers of bush peppergrass are insect-pollinated, and it is known that some other mustards are important sources of nectar and pollen for bees, though apparently the bush peppergrass has received no scientific attention on this point. Cattle appear to occasionally browse on the bush peppergrass, but there seem to be no studies of its palatability. In short, this is another of our desert shrubs about which we know the name but not much else.

ERICACEAE
HEATH FAMILY

Greenleaf Manzanita
Arctostaphylos patula

NEARLY AS symbolic of the West as the sagebrush is the manzanita. The eastern mountains from New England to the Smokies have their characteristic rhododendrons, mountain laurels, and azaleas. Their counterparts in the West, which also happen to belong to the same family, are the various species of manzanita—dominating much of our so-called chaparral vegetation from Canada to Mexico. Within the Great Basin, manzanita generally occurs above sagebrush, where temperatures are cooler and evaporation less. The shiny, crooked, reddish brown stems and leathery, smooth-edged leaves are so distinctive that anyone can learn to recognize this genus after only a brief initial acquaintance.

Nearly everyone is familiar with the dreadful annual fires which sweep through much of the chaparral vegetation, particularly in California. Regrettable as these fires are for humans, they are perfectly natural events in chaparral vegetation, so much so that many of the anatomical and developmental processes of these plants are not only resistant to fire but dependent upon it. Many species produce seeds which will not germinate until exposed to the heat of a fire. On the other hand, about a quarter of the California species of manzanita will resprout after a fire from large, underground root crowns or burls. Frequently, the new growth will be evident within a few weeks.

R. E. Fulton and F. L. Carpenter carried out an intriguing study on two species of manzanita near Riverside, California. One species was a seeder—that is, it depended on seeds to recolonize an area after a burn—while the other species produced root burls and was able to resprout after its aboveground parts were destroyed. Fulton and Carpenter found that the seeder produced more flowers and more nectar and had twice the number of insect visitors than the resprouter. Additionally, the seeder had flowers which were self-compatible; that is, fertilization could be accomplished by pollen from the same plant. The resprouter, however, was not self-compatible. Apparently, in the resprouting species, a lot of energy goes into burl production rather than into flowers and seeds. Obviously, this means that a resprouter can more rapidly reclaim an area than a seeder can. But a price is paid in terms of reduced variability in the population. Vegetative reproduction of this kind means that no reassortment of genes has taken place such as occurs during seed production. In the long run, this is a liability because there is then less probability that some individuals, at least, will have the requisites to survive the inevitable environmental changes.

Fulton and Carpenter, however, suggest that the self-incompatibility of the resprouters insures greater genetic variability and thus more adaptable plants. Self-compatible plants will, if self-pollinated, produce less variation among their offspring. Also, fewer flowers means that pollinators will have to visit other individuals of the same and adjacent species of manzanita to get sufficient nectar and pollen. This would encourage greater variability in the resprouter species than would otherwise be the case. Fulton and Carpenter think that the reseeder forms are more advanced on the evolutionary scale, since these species would have an increased frequency of variation and a greater magnitude of variation upon which the forces of natural selection could act.

Two other workers, J. E. Keeley and P. H. Zedler, proposed that the frequency of fires may select for either resprouting or reseeding species. Very frequent fires would favor the resprouters, whereas longer cycles of ninety to one hundred years would encourage the reseeders. Contrary to what one might assume, the ability to resprout from roots or root crowns is probably a very old trait, and the loss of this ability by the reseeder manzanitas is a much more recent evolutionary trend in the group. Evolutionary adaptations such as these must have developed over many thousands of years, indicating that fire has been a pervasive and determining factor in the West for a very long time. In fact, were it not for fires, much of the chaparral vegetation would eventually be replaced by coniferous or other vegetation over most of its range. This is but one more example of the profound plasticity and adaptability of plants in the harsh environments of our desert and mountain West.

Greenleaf Manzanita

The greenleaf manzanita, as its common name implies, has bright green or yellow-green leaves, unlike the dull green or whitish leaves characteristic of most other species. They are nearly round and from 2.5 to 4.5 centimeters long. Although the leaves are perfectly hairless, the young twigs have a fine pubescence. Gaius Shaver studied the leaf angles and amount of light absorbed by seven species of manzanita in the Sierra Nevada and along the Washington and Oregon coasts. He found that the high-elevation pine-mat manzanita, *A. nevadensis*, had many fewer vertical or near vertical leaves than did the two species at lowest elevations in the Sierras. The greenleaf manzanita, which occupies middle elevations, was intermediate in its leaf angle distribution. The theory here is that under warmer conditions, with intense solar illumination, leaves will be oriented in a fashion that will tend to reduce the amount of light and heat they receive. Under shaded conditions with lower temperatures, we might expect to find that most leaves were horizontal. It would be interesting to know to what extent this orientation is under environmental control—do greenleaf manzanitas in the shade have more horizontal leaves? I suspect that they would, based upon what we know of other plants.

Older branches of the greenleaf manzanita have the smooth, shiny, reddish brown bark so characteristic of the group. Occasionally in the greenleaf manzanita, but more frequently in certain other California species, a phenomenon occurs known as bark striping. These are zones where the bark has disappeared, with the result that the living bark is restricted to narrow "stripes" separated by zones of dead and bleached wood. As such stems mature, they tend to become twisted and gnarled. Craig Davis studied bark striping in a number of manzanitas and found that it could originate in several ways. Drying out of some roots will lead to the death of those branches tied to them physiologically for their supply of water. This would also be expected to cause some of the bark to die. Fire, of course, might be expected to destroy some bark on individuals not completely consumed. Another feature to which Davis attributes much significance is the intolerance of manzanitas to shade. Shaded leaves soon die along with their associated leaves and branches. He found that the extent of bark striping was directly related to the extent of shading. This appears to be an adaptation to the frequent fires, high temperatures, and infrequent moisture of the summer. Bark striping as a result of intolerance to shade reduces water loss from the plant by insuring that only the adequately illuminated leaves and their supporting tissues are kept alive. As Davis says, "The net result of this growth pattern is a vine-like manzanita shrub supported and held aloft by its own dead remains."

The pinkish flowers, borne in nodding terminal clusters, appear during May and June. Each flower is about 5 millimeters long, and its five petals are fused into an urn-shaped structure. This is a flower form common to many members of the heath family, to which the manzanita belongs. There are eight or ten stamens enclosed in the flower, each having a pollen-bearing anther with two bristles or awns bent down toward the base of the flower. The dark brown or black, berrylike fruits have a thin, mealy pulp enclosing several nutlets.

The number of flowers and their consequent fruits produced each year by manzanita will vary considerably. Keeley studied the possible causes of this variation and concluded that it was tied to the amount of precipitation which occurred during the previous year. More moisture meant that the plants were able to carry on more photosynthesis. This in turn resulted in the formation of a greater number of flower buds in the late spring, when they normally develop. Flower buds in manzanita form about a year prior to the time that they mature. They then remain dormant during the summer and the following winter, until flowering occurs in the spring. Fruit formation appeared also to be dependent on food stored during the previous year.

Greenleaf manzanita is the most common manzanita in the Great Basin and in much of the Sierra Nevada, and over much of its range it is the only manzanita. It can be found from pinyon-juniper woodlands up to and above elevations of 9,000 feet. Old burns will typically develop very dense stands of this manzanita. Aside from its ability to sprout from root burls, branches which become weighted down by snow will readily produce roots and thus cause an even more rapid colonization to occur. Eventually, conifers will invade manzanita stands and shade them out, provided no additional fires occur.

Although its berries and young foliage seem to be of some value for wildlife, greenleaf manzanita is considered virtually worthless for livestock. How worthless was shown vividly by an experiment conducted in the Lassen National Forest in California. J. H. Hatton reported that, although goats penned in a manzanita area girdled and killed many of the plants, the goats nearly starved to death during the second year. Manzanita has been made minor use of by humans. Great Basin Indians reportedly ate the fruits occasionally and made an extract from the leaves for use as a diuretic. The berries can be made into a jelly, and the seeds can be ground into flour.

The greenleaf manzanita ranges north into Oregon, east through Utah to Colorado, and south to the higher elevations above 8,000 feet in Arizona. It occasionally hybridizes with other manzanita species. Above Spooner Summit, west of Carson City, Nevada, a small colony of hybrids shows varying degrees of intergradation between the greenleaf manzanita and the pinemat manzanita characteristic of higher elevations in the Sierra Nevada. In locations where the greenleaf distribution overlaps that of several lower-elevation forms, there have also been reports of hybrid intergrades.

The genus name *Arctostaphylos* comes from two Greek words, *arktos*, bear, and *staphule*, grape. One species of manzanita, in fact, is known as *uva-ursi* or bearberry—*ursa* is the Latin word for bear. The bearberry is, incidentally, the only manzanita that has a range beyond the confines of North and Central America. It has a circumpolar distribution in the northern hemisphere, being found throughout Canada, northern Europe, and northern Asia. The common name of manzanita means little apple in Spanish, in reference to the shape of the fruits. The species name *patula* is derived from a Latin word meaning wide-spreading. As we noted earlier, manzanitas belong to the heath family or Ericaceae, discussed in our treatment of the western blueberry.

Western Blueberry
Vaccinium uliginosum ssp. *occidentale*

Her ankles were black from the dirt of the fields, and her hands were midnight-blue from the wax of the berries. In her home, she served each of her visitors a blueberry that was the size of a baseball, as they recall it, heaped over with sugar and resting in a pool of cream. Then she asked them to consider planting blueberry bushes along the Garden State Parkway.
—John McPhee, THE PINE BARRENS

EVEN IN THE New Jersey pine barrens, blueberries never really get larger than grapes, much less the size of baseballs, but the memory of their delectable taste undoubtedly caused them to seem bigger than they were! For the most part, our soils in the Great Basin are far too alkaline to allow blueberries to grow, let alone flourish. Consequently, our western blueberry is found in the higher mountains within the lodgepole pine and subalpine forest zones, where acid soils may be expected. Typically, it is found along streams or in wet meadows. A small shrub generally about 30 centimeters high, it may on occasion become nearly a meter high in favorable locations. The berries are blue-black and about 6 millimeters in diameter. While they are sweet, they are not equal to their eastern counterparts in flavor. The bluish color is due to a waxy bloom on the fruits, like that of blue grapes.

The leaves on the western blueberry are 1 to 2 centimeters long, elliptical or oval in shape, and lacking any teeth along the edge. They are pale green above and even paler beneath. In the fall the leaves turn yellow or reddish, in concert with the trembling aspens, before they wither and drop. The white or pinkish flowers are produced in clusters of two to four at the upper ends of the branches during June and July. The pollen-producing portion, or anther, of each stamen has an intriguing appearance when viewed under a

Western Blueberry

hand lens. The top portion of each anther is drawn out into two narrow tubules, and adjacent to these are two hornlike protuberances. The tubules are open at the upper end, and the sticky pollen comes out through these tubules very much like toothpaste out of a tube as the anthers mature. The sticky pollen readily adheres to any pollinating insects. Anthers which open at the top by slits or pores are the rule for members of the heath family, to which the blueberry belongs.

Blueberries have the distinction of being one of the few fruits introduced into cultivation in this century, although both settlers and Indians for centuries before made extensive use of wild blueberries. Early in this century, Elizabeth C. White of Whitesbog, New Jersey, the daughter of a cranberry grower, became interested in developing the potential of the highbush blueberry and offered prizes for the biggest fruit. Frederick V. Coville, a U.S. Department of Agriculture horticulturist who had previously written about the possibility of producing a superior blueberry, learned of her interest and offered to work with her. Coville and White began to grow a number of the more promising forms on her property. The early varieties were forms selected from the wild, but soon these were replaced with superior types developed through extensive hybridization programs.

The first variety which White and Coville released commercially was called the Pioneer. There are now hundreds of varieties produced by Coville's successors since his death in 1937. As important as the production of new varieties was, the development of an entire system of agriculture was needed to grow the plants, which are quite different in their requirements from most other cultivated fruits. Blueberries require an acid, well-drained sandy loam with a pH between 4.2 and 5.2, with the water table preferably not too far from the surface. In nature, pruning is accomplished by the nearly annual light fires which were common at one time in the blueberry communities. Coville found that the maximum number of fruits were produced when the greatest number of two- and three-year-old branches occurred on the plant. Consequently, pruning practices ideally have tended toward this goal—in imitation of nature, as it were. Interestingly enough, the severity of pruning will affect both the number and the size of the berries, as well as the time of ripening. A light pruning will result in more and smaller berries which ripen later. Coville found that cross-pollination with a different variety resulted in better fruit production than was the case with self-pollination.

The genus *Vaccinium* is now considered to include both the blueberries and the huckleberries, though at one time the latter were separated into the genus *Gaylussacia*. Huckleberries have ten or so bony nutlets in the fruits. Each nutlet is really comparable, though it is much smaller, to a peach pit. Blueberries, on the other hand, have fifty or more very small seeds, not enclosed in any hard, bonelike covering. In Europe members of this genus are referred to as bilberries or whortleberries, and records of their use for human consumption are found throughout historic times. *Vaccinium* was the Latin name for bilberry. The subspecies name *occidentale* means western, while the species name means wet or marshy in reference to its habitat. There are several other species of blueberry common in the Northwest, but only *occidentale* gets well within the Great Basin.

The young twigs and leaves, considered to be good browse for deer, are variously regarded as good to poor for cattle and sheep. The western blueberry ranges from the Sierra Nevada in California through the higher mountains of Nevada east to Utah and Montana. To the north it reaches British Columbia.

The heath family, known as the Ericaceae, is worldwide in distribution and contains around one hundred genera and three thousand species. The Himalayas, South Africa, and New Guinea contain particularly large concentrations of species. Azaleas, rhododendrons, and the manzanita of the Sierra Nevada are members of the same family.

GROSSULARIACEAE
CURRANT & GOOSEBERRY FAMILY

Western Golden Currant
Ribes aureum

THE WESTERN GOLDEN CURRANT is far and away the most attractive of our native species of currant or gooseberry. It is highly regarded, not only for its showy yellow flowers produced in spring but also for its relatively sweet, juicy orange berries. Some forms produce black or red berries. Howard McMinn in his book *An Illustrated Manual of California Shrubs* tells of some golden currants in the trial garden at Mills College, California, which produced all black fruits early in the season and all orange fruits later on. The parent plant was said to have produced red, black, or orange fruits. This striking color variation is most probably due to a simple inheritance pattern, though no one has yet worked out the precise genetic mechanism involved.

An edible variety known as the Crandall as well as several others have been developed, although they are rarely grown. Most white and red currants in cultivation belong to the Eurasian species, *R. sativum*. The commercial black currant, *R. nigrum*, is little grown in this country.

The western golden currant is widely distributed in the Great Basin and in the West generally. It is sometimes quite abundant along irrigation ditches and the floodplains of streams. Although the wax currant is occasionally cultivated for ornament, the western golden currant is the most widely grown of our native Great Basin species. In suitable locations it will grow taller than 2 meters, and horticulturists have found that it is easily propagated from cuttings. The branches, which produce a gray or brown bark, bear three- or five-lobed leaves with conspicuous veins. They vary in length from 1 to 3 centimeters. Both the stems and the leaves are without pubescence, though rarely the leaves may show some sign of it.

In our discussion of the wax currant, we point out that the variation in the nature of the floral tube in currants and gooseberries extends from forms

Western Golden Current

with broad, open, cuplike flowers to varieties with narrow, cylindrical flowers. The western golden currant is especially notable in that it has one of the longest and narrowest tubular flowers among our native forms. The tubular, yellow portion is about 1 centimeter long. At the mouth of the tube are five small, yellow petals about 3 millimeters long, alternating with the larger sepal lobes, which are about half or a little less than half the length of the tube. In some forms, the petals of older flowers become somewhat reddish. Because of the narrow, tubular nature of the flower, the number of different insect visitors is restricted compared to species with a more open form. Some insects, however, are able to circumvent this limitation by simply boring a hole through the base of the flower to get at the nectar.

The western golden currant has a wide range over the West, extending from the Los Angeles area north to British Columbia and east to the Rocky Mountains. McMinn notes that in southern California and the south Coast ranges "the flowers lack the spicy odor of the plants east of the Sierra Nevada and in the Rocky Mountains." A form known as variety *gracillimum*, which grows near the coast of California, has flowers with a tubular portion relatively longer compared to the calyx lobes than is true of the typical variety. But this is at best a minor difference, not recognized as significant by some botanists. The original specimen of *R. aureum* on which the description is based, the so-called type specimen, was collected by Meriwether Lewis in the Rocky Mountains during the Lewis and Clark Expedition of 1804 to 1806.

The common name of currant has a curious origin. Corinth, Greece, was the main port for the export of seedless raisins during medieval times. In France, these became known as the raisins of Courantz, and only a minor linguistic change was required to arrive at "currant." The name was applied to our modern currant because black currants, especially, resemble the "raisins of Corinth." The scientific name *Ribes* has a similar convoluted origin. It appears to be derived from the Arabic *ribas*, which was applied to a medicinal species of rhubarb. When the Moors invaded Spain in the eighth century, they gave the name to a plant which they perceived as having similar properties, since there was no native rhubarb in Spain. Apparently the word meant a plant with acid juice. In Italy, one species is still known by the common name of *ribas*.

Wax Currant
Ribes cereum

DRY, OPEN SLOPES throughout the Great Basin favor the development of a shrub community which frequently includes one or more species of currant or gooseberry. One of the most widespread of these is the wax currant, which ranges from British Columbia to southern California and east to the Rocky Mountains. From the lower mountain slopes it frequently extends upward to the alpine tundra zone.

Typically, the wax currant is about a meter tall. It is easily distinguished by its small leaves, generally about 25 millimeters wide, which are roughly circular in outline, with fine teeth along the edges and a tendency to be three- or five-lobed. In addition, the stems, unlike those of many members of the genus, totally lack any spines. On drier sites, the leaves are noticeably smaller, while currants on moist, shaded sites may have leaves 50 millimeters across. The leaves tend to be clustered at the ends of spurlike branches. Within the Great Basin, the chokecherry and serviceberry are common associates, although farther to the northwest the bitterbrush is a frequent companion. Not infrequently, this currant and the plateau gooseberry appear to be growing on mounds of soil. James Young and his associates found that these so-called ribes mounds had simply accumulated erosion products, plant debris, etc., and were not indicators of old erosion surfaces, as some had supposed.

The white to pink, drooping clusters of flowers are produced in the late spring and early summer; they are tubular and about a centimeter long. The tube, really the fused base of the calyx (sepals) and corolla (petals), is perched on top of that portion of the pistil which contains the developing seeds. The free portions of the petals are very tiny, white structures between the triangular lobes of the sepals at the top of the floral tube. This kind of

Wax Currant

fusion in flower parts is considered an advanced feature, one which appears to have been a major evolutionary step toward more efficient pollination.

The flowers are followed by red berries, about 12 millimeters in diameter, which ripen in the fall. These are not particularly flavorful, although like the fruits of all members of this genus they are edible and may be used in jams and preserves. Pemmican, a mixture of dried meat and berries, was prepared by some western Indian tribes from various native currants or gooseberries. The wax currant is considered only fair to poor browse for cattle and deer, but it is nevertheless important on ranges where little else may be available.

The genus *Ribes* contains about 150 species, with its main center of concentration being in North America, although it also occurs in Asia, Europe, and South America. The spiny members of the genus are generally called gooseberries and may have fruits which are smooth or spiny. At one time they were separated into the genus *Grossularia*. However, this is now considered an artificial distinction. The primary reason for this conclusion is that fruits, unlike flowers, are considered to be very plastic in their evolution. Not infrequently, in many genera, one will find that some species produce dry capsules, others fleshy berries, while still others develop complex aggregate fruits composed of many ripened pistils adhering together. The conservative nature of flower evolution may be due to the complexity of the processes of pollination and fertilization and the probability that any tinkering with this chain of events by random mutations in nature is very likely to throw the whole mechanism awry. On the other hand, development of fruits is a simpler affair, and one might expect that plants would be able to evolve various types of fruits in short order as adaptations to the presence or absence of particular dispersal agents.

Currants belong to the family Grossulariaceae, which is closely related to the saxifrages and hydrangeas; in fact, they are considered by some to belong to the same family. All of the currants and gooseberries have the sepals and petals fused to form either a broad, open cup or a narrower tube of varying dimensions, depending upon the species. The five stamens are situated near the upper end of the tube. Bees will visit the broad, open types as well as the narrower types for pollen, but nectar from those latter forms which are especially long and narrow can be obtained only by hummingbirds and long-tongued bees. Like many other hummingbird flowers, the longer tubular types may represent an evolutionary adaptation to the hummingbird beak. Most bird-pollinated flowers are found in the tropics, but, wherever hummingbirds occur in the temperate zones, one may also expect to find some flowers which have coevolved with them.

One of the most prolific and tireless observers of the process of pollination was Hermann Muller, who tabulated his observations of thousands of plants in a work published in Germany in 1873. This was later brought up to date in a three-volume work by Paul Knuth, the *Handbook of Flower Pollination*, which has remained a classic and an invaluable reference since its publication in English in 1906. Muller observed the visitors to a number of currants and gooseberries and pointed out that those with narrow tubes and pendulous flowers discouraged pollination by flies. Pendulous flowers, along with the curled-back sepals of many ribes species, also deter entrance by ants. Ants and flies would be such inefficient pollinating agents that we can suppose evolution to have favored those species which were able to evolve structures to guard against these freeloaders.

The genus name *Ribes* comes from an Arabic name for a plant with acid berries, which, however, was really a species of rhubarb found in southwestern Asia. Medieval herbalists mistakenly applied the name to the red currant of Europe. The species name *cereum* means waxy.

Plateau Gooseberry
Ribes velutinum

OF THE SEVERAL gooseberries and currants common in the Great Basin, the plateau gooseberry appears to be capable of growing on the driest sites. It is common on dry mountain slopes, especially in sagebrush areas. It ranges north to Oregon, east to Utah and Arizona, and, on the eastern side of the Sierra Nevada, south as far as the White Mountains.

The plateau gooseberry is a rather rigidly branched shrub, sometimes as much as 2 meters high, though more typically it is 1 meter or less. This gooseberry is easily recognized by its small (1 to 2 centimeters), deeply cleft, five-lobed leaves, which are roughly circular in outline, combined with the single large spine at each node. Unlike certain other spiny gooseberries, there are no spines on the internodes.

In the spring this gooseberry produces small, inconspicuous flowers—usually whitish or yellowish—with several grouped together at the nodes. The fruit is a small, dark purple, and rather dry berry. The leaves, young stems, and fruits are covered with a soft pubescence. There is a variety located outside the Great Basin known as *glanduliferum* because its pubescence includes hairs with swollen, glandular tips. This form ranges from northern California to the San Gabriel and Kingston mountains at the southern end of the state.

In common with other members of the genus, the plateau gooseberry provides only fair to poor forage for livestock. Because of the spines on this species, it seems to be avoided by cattle. However, various species of wildlife from deer to chipmunks and such birds as scrub jays and magpies apparently make good use of gooseberries and currants.

Ribes is considered a liability by eastern foresters, since several species serve as the alternate host for the white-pine blister rust. In order for this

Plateau Gooseberry

parasitic fungus to complete its life cycle, currants or gooseberries must be nearby for certain stages to grow on. While the disease is not a problem in the Great Basin, it was very significant in the eastern United States, since it affected an important timber tree. The most feasible control measure, practiced since the early 1900s, has involved the eradication of currants and gooseberries, both native and introduced, with the result that in many areas of the East *Ribes* is a relatively rare shrub where it was once common.

Some species of ribes are known to accumulate aluminum. Why this should be so is uncertain, for as far as most plants are concerned this element is toxic in soluble form and, so far as is known, is not an essential element for growth or development. Clay, of course, contains an abundance of aluminum, but this is in an insoluble form and so is not harmful to plants. Aluminum apparently interferes with the uptake of calcium and iron and ties up phosphorus in roots, preventing its transport to the rest of the plant.

Characteristics of the genus *Ribes* are discussed under the wax currant. The species name *velutinum*, from the Latin, means covered with a silky pubescence. The family Grossulariaceae appears to be relatively old, since fossil wood similar to that of present-day *Ribes* has been found in deposits belonging to the Upper Cretaceous period in California. This would place the age of the gooseberry family, based on the wood of these specimens of *Riboidoxylon*, at between 70 and 80 million years. Fossil leaves resembling those of *Ribes* have been found at a number of sites of a later age. One estimate states that the family possesses some 350 species in 25 genera. Its distribution includes all of the continents except Antarctica.

ROSACEAE
ROSE FAMILY

Western Serviceberry
Amelanchier alnifolia

OUR WESTERN SERVICEBERRY is consistent only in its extremely variable nature, perhaps more so than any other shrub we have. Its variability includes not only a number of differences in leaf and stem form but a large array of physiological types as well, which appear to be adapted to an assortment of habitats. Western serviceberry can be found in habitats ranging from open, dry, and rocky slopes to the deep shade of coniferous forests. It is found from sea level in Washington State to well over 9,000 feet in the Sierra Nevada of California. In shaded locations it is a small, sprawling shrub, sometimes only a few centimeters high. In better-lighted locations, western serviceberry may get to be over 2 meters tall. Sometimes a few erect branches begin to grow near the center of what was a low shrub and, in this way, convert it within a few years into a small tree.

Because of this variability, some botanists have characterized numerous forms of western serviceberry by a series of species and varietal names. One form, known as *A. pumila*, has leaves which at the most are only slightly pubescent at maturity; it is especially characteristic of damp woods. Another species, *A. pallida*, has leaves with a fine pubescence, particularly on the lower surface, and seven to nine pairs of lateral veins joining the midrib. This species is commonly found on dry, rocky slopes. Many of the specimens from the western Great Basin appear to belong to this form. Also common in dry habitats throughout the Great Basin is another species, *A. utahensis*. It is similar to the preceding species but has somewhat smaller flowers and leaves with nine to thirteen pairs of lateral veins. However, many botanists regard these names with some trepidation, because of the existence of many examples which are obviously intergrades. And the existence of numerous intergrading forms between two "species" is considered by some authorities

Western Serviceberry

to indicate that they are not really distinct species but only variations of one species. That is, species are supposed to maintain their integrity and not hybridize so readily (though many do).

All taxonomists know, or think they know, the rules for defining a species. I recall a three-day seminar at the Missouri Botanical Garden some years ago dealing with the topic, "What is a species?" There were nearly as many different opinions as there were speakers. One of the speakers, only half joking, suggested that a species was whatever you consider to be a species. The problem basically involves getting other experts to agree with your characterization of any particular species. To a great extent, species recognition is as much an art as it is a science, at least in some genera. It is conceivable, of course, that the western serviceberry is now experiencing a rapid evolution which in the course of a few thousand or tens of thousands of years will result in many new species. Because so little change is apparent in a human lifetime, we sometimes forget not only that evolution is a process that has operated over the last 3 billion years to create the myriad life forms which have become extinct as well as those which now exist on the earth, but that such evolution continues unabated today and will perpetually generate new species as long as life exists. It follows, then, that we will occasionally see this process in operation to an extent that makes it difficult to categorize species in some genera.

The leaves on the western serviceberry, simple and oval in shape, are about 2 to 4.5 centimeters long. They are prominently veined, and usually there are teeth along the leaf edge above its middle, although occasionally they extend to the base or are lacking altogether. The white flowers are usually produced several together on a central stalk. Like many other members of the rose family, to which the serviceberry belongs, there are five sepals and five petals. The sepals are fused at the base, along with the apex of the flower stalk, to the ovary of the pistil—the ovary is that part of the pistil which contains the ovules that later develop into seeds. Flowers built along these lines appear to have the sepals, petals, and stamens perched on top of the ovary and are said to have an inferior ovary. We now consider such flowers to be advanced on the evolutionary scale. Examine an apple flower, and it will be seen to have a structure very similar to that of the serviceberry. Both the serviceberry and the apple are among the more "progressive" members of the rose family.

The flowering period in May and June is followed fairly soon by the development of bluish or black berries (which, technically, are more like apples than true berries). The early maturation of the fruits accounts for the name Juneberry in some locales. In the East, *Amelanchier* sometimes goes by the name of shadbush or shadblow because its blooming period coincides with the spawning migration of shad. Other names in the West are pigeonberry, alderleaf sarvisberry, and saskatoon. The name serviceberry comes from the similarity of its fruits to those of the European service tree, *Sorbus domestica*.

The western serviceberry has a range extending north to Alaska and eastward to Michigan. Serviceberry is considered an important and palatable browse plant for both cattle and sheep, and it is said to be a valuable food for deer and elk. The berries are sweet and can be eaten either raw or cooked in various ways, typically as jams or jellies. Our western serviceberry has fruits which are superior to those of the other species found in North America. Some varieties have been introduced into cultivation and have been given horticultural names. Various tribes made pemmican from serviceberry, and the Ute Indians are said to have combined ground-up grasshoppers and Juneberries for a nutritious repast!

There are about twenty species of *Amelanchier* in the temperate areas of the northern hemisphere. The genus name comes from the Savoy name for the medlar, *Mespilus germanica*, a tree of southern Europe and also a member of the rose family. The species name *alnifolia* is an obvious reference to the resemblance between the serviceberry's leaves and those of the alder.

Littleleaf Mountain-Mahogany
Cercocarpus intricatus

ALTHOUGH CONSIDERED by some to be only a variant of the curlleaf mountain-mahogany, *C. ledifolius*, a species which is covered in Lanner's *Trees of the Great Basin* in this series, the littleleaf mountain-mahogany in its usual form is nevertheless very distinctive and easy to separate from its larger and commoner relative. Typically only 1 or 2 meters tall, the littleleaf is aptly named, for its leaves are only about a centimeter long, whereas those of the curlleaf range up to 2.5 or 3 centimeters in length. In both, the tough, leathery, evergreen leaves are similarly shaped, with the smooth edges curled toward the white-pubescent underside.

Absent from the western Great Basin, littleleaf has a more southerly distribution than curlleaf—it ranges from the White Mountains of California through southern Nevada to White Pine and Elko counties in eastern Nevada, then through Utah to southwestern Colorado and Arizona. Why it should be absent from the central Great Basin, where the curlleaf is common, is unknown. Possibly its habitat requirements, which have not been studied, are slightly different. Ivar Tidestrom in his *Flora of Utah and Nevada*, published in 1925, mentions that the littleleaf intergrades with curlleaf, even though he retains the former as a distinct species.

We might legitimately ask at this point, When is a form a distinct species or only a variety or a subspecies of another species? Unfortunately, there is no clear-cut and solely objective way to answer this question. Any species, in order to be so considered, must differ in at least several visible ways from other, similar species. If only a single difference occurs—if the plant has white flowers rather than blue, for example—then nearly everyone agrees that such a form cannot be considered even a variety or a subspecies, much less a distinct species. Modern systematics recognizes the importance of

Littleleaf Mountain-Mahogany

present-day genetic theory, which says that simple, one-character variations could easily be the result of a single gene difference. This difference may come about through a simple mutation, though there are other ways in which such differences could arise. At any rate, single gene differences do not make for a new species.

In the present instance, where there are several obvious physical differences, the situation is complicated by the existence of a number of intergrades. These might be regarded as the result of hybridization between two species or, alternatively, as simple variations within a single species. To a great extent, more than most botanists like to admit, the issue is decided by whoever happens to be the latest expert on the group in question. If this expert tends philosophically to be a splitter, then the two forms will probably be considered separate and distinct species; however, if our expert finds the outlook of a lumper to be more satisfying, then the littleleaf mountain-mahogany may be viewed as only a variant of the curlleaf, and the species *intricatus* will be sunk and maintained only as a variety or a subspecies. The next expert to come along, on the basis of additional evidence and/or philosophical inclinations, might reverse the situation once again. If all this seems depressing, it should not be, for it must be remembered that the species concept is a human invention and, consequently, only an imperfect description of nature.

Typically, littleleaf occurs in dry, rocky desert ranges, frequently mixed in with pinyon-juniper. Unlike the curlleaf, it rarely occurs as a continuous stand over a large area. The small, inconspicuous flowers are produced from April to June, depending on the elevation, and are borne one or two together on very short stalks among the leaves on the younger branches. Each flower, less than a centimeter long, is funnel-shaped with five lobes along the upper edge. There are no petals, and the inside of the funnel has numerous stamens perched on it. At the bottom of the funnel is a single pistil with a long style perched on top of the ovary, which in turn contains a single seed. After the flowers have been pollinated, the style becomes hairy and twisted and elongates to between 2.5 and 4.5 centimeters. At this stage, during the summer, the fruits are so abundant on some shrubs that from a distance they appear to be coated with frost. These conspicuous fruits are not immediately dispersed and may persist well into winter. The hairy styles attached to the seeds serve to carry them on the wind some distance away from the parent plant.

The wood of mountain-mahogany is extremely hard; in some areas it has been overharvested because of its desirability as firewood. Some Great Basin Indians used curlleaf wood for bow construction, though the littleleaf probably never got large enough for that purpose. More important, mountain-mahogany was regarded as a valuable medicinal source. The bark was dried and used as a remedy for tuberculosis, as well as for colds and other respiratory problems. Powdered bark was used as a remedy for sores and various wounds. A tea prepared from the bark or leaves was also widely revered as a remedy for heart disease and other complaints, ranging from stomachaches to venereal disease. Littleleaf as well as curlleaf mountain-mahogany is considered good winter browse for deer and elk. Cattle and sheep will feed on it to some extent during the cold season, but neither is generally regarded as a shrub of major significance for livestock.

Two other mountain-mahoganies are to be encountered on the edges of the Great Basin. One of these, the western mountain-mahogany, *C. betuloides*, is native to the western slope of the Sierra Nevada in California and Oregon. It also ranges south into Arizona and Mexico. There are reports of occasional examples on the eastern slope of the Sierras in Nevada, though it remains an extremely rare shrub in that area. The western mountain-mahogany is easily identified by its wedge-shaped, deciduous leaves, toothed —bove the middle and 1 to 3 centimeters long, dark green and smooth on the upper side. The leaves resemble those of the western serviceberry, except for their smaller size. Unlike its leathery-leaved relative, western mountain-mahogany is considered an important browse species for livestock and deer. On the eastern edge of the Basin in Utah and in extreme eastern Nevada is another species, the alderleaf mountain-mahogany, *C. montanus*. It has similarly shaped deciduous leaves which differ most obviously from those of the western mountain-mahogany in being sparsely gray-pubescent on the upper surface. Like the western species, the alderleaf is considered an important browse plant.

The genus name *Cercocarpus* comes from two Greek words, *kerkis* meaning shuttle and *karpos* meaning fruit, in an allusion to the resemblance between the fruit with its long, twisted tail and the instrument used in weaving. The species name *intricatus* is from the Latin and refers to the characteristic entangled branches.

The mountain-mahoganies belong to the rose family. For a few of the characteristics of this family, the reader should consult the discussions on desert peach, western serviceberry, and blackbrush. Altogether there are about twenty species of *Cercocarpus*, all of them in North America, mostly in the western United States.

Fern Bush
Chamaebatiaria millefolium

EITHER COMMON NAME, fern bush or desert sweet, is certainly appropriate for *Chamaebatiaria*, since its leaves are like minute fern fronds and the whole plant has a pleasant fragrance. The leaves, which are only 2 to 5 millimeters long, are clustered near the ends of the twigs in whorls which resemble a miniature cycad. The symmetrical design of the fern bush is in harmony with the rocky habitat in which it most commonly grows, frequently in the company of juniper and pinyon pine. Usually densely branched, it attains a height of 1 to 2 meters.

During the summer, the younger twigs bear elongated clusters of white flowers about 15 millimeters across. Each flower has the typical form of the more primitive members of the rose family to which it belongs, inasmuch as there are five sepals, five petals, numerous stamens, and five pistils, with the latter more or less fused together within a small, cuplike structure. Each of these pistils eventually matures into a dry, brown pod containing several seeds.

The fern bush skirts most of the Great Basin, since it has a distribution along the eastern side of the Sierra Nevada north into Oregon and Idaho and then south into Utah, eastern Nevada, Wyoming, and Arizona. In northern Nevada it can be found as far west as the area around Ely, though it is more common in southern Nevada in the mountains. It is common in the raw lava fields of northeastern California. Although it might appear to be eminently suitable for cultivation, it remains as infrequent in gardens as it was in 1900, when Liberty Hyde Bailey noted in his *Cyclopedia of American Horticulture* that it was "rarely cultivated," even though it was hardy as far north as Massachusetts, and "likely to be killed by too much moisture during

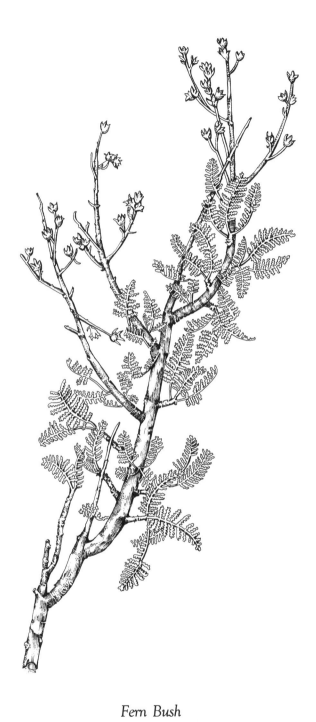

Fern Bush

the winter." The most recent edition of the standard Sunset "New Western Garden Book" makes no mention of the fern bush.

Indians in the Great Basin made use of a tea prepared from fern bush leaves as a treatment for cramps and stomachaches. Sheep and deer, but not cattle, are said to browse on fern bush occasionally. In any case, the fern bush is not really abundant anywhere and cannot be regarded as an important food plant.

The affinities of the fern bush are well illustrated by the name changes it has undergone since it was originally described in 1857, from material collected in the Williams Mountains in Arizona. Then it was considered to be a *Spiraea*, because of the resemblance of its flowers to those of other members of that genus. Bailey lists it under *Sorbaria* because of its dissected leaves. It was once placed in the very similarly named genus *Chamaebatia* as well as one other genus. We hope that it has found its final resting place, though botanists, like other scientists, never seem to be able to make up their minds. Of course, botany like any other field is always progressing and developing new insights, and we could validly say that any intellectual enterprise which finally and conclusively makes up its mind has in all probability become stultified, if not moribund!

Chamaebatiaria gets its name from its close resemblance to *Chamaebatia*, otherwise known as Sierra mountain misery, a plant found in California but not in the Great Basin. The latter is derived from the Greek *chamae*, which means low, and *batos* signifying bramble, referring to the bramblelike flowers. The species name *millefolium* means many leaves, but in this case it is more obviously a reference to the resemblance between the fern bush leaf and that of the common weedy yarrow or milfoil.

Green Ephedra. *Rick Stetter.*

Dwarf Juniper. *Rick Stetter.*

Nevada Ephedra, male cones. *Rick Stetter.*

Four-winged Saltbush, male flower. *Rick Stetter.*

Iodine Bush, winter aspect. *Hugh Mozingo.*

Four-winged Saltbush. *Stephen Trimble.*

Four-winged Saltbush, with frost. *Stephen Trimble.*

Four-winged Saltbush, winged fruits.
Hugh Mozingo.

Torrey Saltbush. *Rick Stetter.*

Winterfat. *Hugh Mozingo.*

Shadscale. *Rick Stetter.*

Big Greasewood, male flowers. *Rick Stetter*.

Bailey's Greasewood, female flowers. *Rick Stetter*.

Heermann's Buckwheat. *Rick Stetter*.

Coyote Willow, female catkins. *Rick Stetter*.

Kearney's Buckwheat. *Christine Stetter.*

Spiny Hopsage. *Rick Stetter.*

Rock Buckwheat. *Hugh Mozingo.*

Bush Chinquapin. *Rick Stetter.*

Wax Currant. *Rick Stetter.*

Tamarisk. *Tony Diebold.*

Purple Sage. *Rick Stetter.*

Red Elderberry. *Rick Stetter.*

Double Honeysuckle. *Rick Stetter.*

Snowberry. *Rick Stetter.*

Big Sagebrush. *John Running.*

Big Sagebrush. *Stephen Trimble.*

Rubber Rabbitbrush. *Stephen Trimble.*

Bud Sagebrush. *Rick Stetter.*

Shortspine Horsebrush. *Stephen Trimble*.

White Burrobush. *Rick Stetter.*

Shockley's Desert Thorn. *Rick Stetter.*

Gray Horsebrush. *Rick Stetter.*

Poison Oak. *Rick Stetter.*

Western Blueberry. *Hugh Mozingo.*

Greenleaf Manzanita. *Rick Stetter.*

Greenleaf Manzanita, fruits. *Rick Stetter.*

Fernbush. *Rick Stetter.*

Bitter Cherry. *Hugh Mozingo.*

Western Serviceberry. *Rick Stetter.*

Littleleaf Mountain-Mahogany. *Rick Stetter.*

Cliffrose. *Rick Stetter.*

Cliffrose, feathery fruits. *Stephen Trimble.*

Desert Peach. *Chris Ross.*

Ocean Spray. *Rick Stetter.*

Wild Rose. *Rick Stetter.*

Bitterbrush. *Hugh Mozingo.*

Wild Rose, fruits. *Rick Stetter.*

Smokebush. *Rick Stetter.*

American Dogwood. *Rick Stetter.*

Silver Buffaloberry. *Rick Stetter.*

Squawbush. *Stephen Trimble.*

Tobacco Brush. *Rick Stetter.*

Prickly Phlox. *Rick Stetter.*

Sierra Coffeeberry. *Hugh Mozingo.*

Spiny Greasebush. *Rick Stetter.*

Blackbrush
Coleogyne ramosissima

THE BLACKBRUSH or, as it is sometimes called, the burrobrush gets its common name from the older stems, which have a rough, black bark. On younger stems the bark is ashy gray. A characteristic feature of the blackbrush is its pattern of very intricate branches, with the branchlets usually two at a node, opposite each other. Each successive pair of branchlets tend to be at right angles to the stems from which they arise. The smaller branchlets end in spines. This combination of characters makes the blackbrush easy to recognize, even from a distance. Ordinarily, the blackbrush varies from a third of a meter to a meter in height, though in rare cases it may reach a height of 2 meters.

The deciduous leaves are small, gray, and narrow, between 5 and 12 millimeters in length, and characteristically clustered at the ends of smaller branchlets. Actually, like the branchlets, they are borne two at each node. They are finely pubescent and have two to four grooves beneath. Under a hand lens the fine hairs on the leaf will be seen to be attached at the middle. The individual flowers—yellow to brownish and about a centimeter wide—are produced from early spring until midsummer. The yellow color is due to the sepals, of which there are four. These are fused at the base and subtended by one or two pairs of greenish, three-lobed bracts. There are no true petals. The numerous stamens are attached to a sheath which surrounds the pistil. This is a unique structure, sometimes called a torus, and in this case it is densely white-pubescent inside. The pistil which the torus encloses contains only one seed at maturity. The style, which appears to arise about two-thirds of the way up the side of the ovary, is densely pubescent at its base. At maturity, the ovary with its single enclosed seed becomes hard and forms a small, brown nutlet or achene.

Desert bighorn sheep and deer apparently make some use of this plant for browse, while ground squirrels and quail collect the achenes. For cattle and horses, however, it is considered to be poor to useless.

Under the auspices of the U.S. National Herbarium at the Smithsonian Institution, Ivar Tidestrom published the *Flora of Utah and Nevada* in 1925. Tidestrom pointed out that the blackbrush within the Great Basin occupied a broad belt between the northern and southern deserts. In the north it is commonly associated with hopsage and may extend into the pinyon-juniper community, while to the south the creosote bush may be a companion. In many instances, blackbrush may form relatively pure stands with only a minority of other shrubs present. It is particularly abundant on gravelly slopes and foothills as well as desert mesas below 6,000 feet. Some authorities consider that blackbrush marks the upper limit of Mohave Desert vegetation. To the west, the blackbrush extends into the Colorado Desert of southern California, while in Utah it is found in the area of the Green and San Juan rivers in Emery and Grand counties. It just barely gets into southwestern Colorado and is common in northern Arizona.

The genus name *Coleogyne* is derived from two Greek words—*koleos*, sheath, and *gune*, ovary—in reference to the torus which surrounds the pistil. The species name *ramosissima* means with many branches. There are no other species in the genus. As a member of the rose family, Rosaceae, blackbrush is a rather atypical representative, since relatively few roses have evolved into desert shrubs. It is also unusual in having four sepals rather than five, as do most other members of the family. Even more unique, however, is the torus or sheath surrounding the pistil. The individual genera in the rose family have flowers in which the sepals, petals, and stamens are variously fused to form a disk or a cup. The apple fruit is considered by some analysts to be such a cup, which has fused with the pistil making up the core of the apple. However, this is not the explanation for the torus of the blackbrush. The answer to this minor but interesting problem remains for some avid student of our desert botany to formulate.

Cliffrose

Cowania mexicana var. *stansburiana*

THE CLIFFROSE, at first glance, somewhat resembles the common bitterbrush. A closer look, however, reveals a number of differences. The 6- to 14-millimeter-long leaves have five to seven narrow lobes, rather than being wedge-shaped and three-lobed at the end as in bitterbrush. The leaves of both species are whitish-pubescent beneath. Additionally, cliffrose leaves are peppered with translucent dots which are particularly apparent when the leaf is held up to the light. Botanists call this a glandular-punctate condition—orange peels show this same feature. On much of the plant, the leaves are crowded at the ends of short twigs. Ordinarily ranging in height from 1 to 4 meters, the cliffrose occasionally develops into a small tree up to 8 meters tall. Older trunks develop a loose, shreddy bark.

When in bloom, the cliffrose is unmistakable, for the 14- to 20-millimeter-broad, cream-colored to sulfur yellow flowers covering the plant are quite conspicuous compared to the smaller, pale yellow flowers of the bitterbrush. The only other shrub with which the cliffrose might be confused is the Apache plume, *Fallugia paradoxa*, to which it is closely related. The Apache plume, however, has larger (25 to 35 millimeters in diameter), pure white flowers borne singly to severally on elongate, erect stalks. It also has a more southerly distribution, essentially outside of the Great Basin proper.

Each of the flowers has five to ten pistils. These mature into dry, hard fruits with long, plumelike tails developed from the styles—from a distance the cliffrose appears as a feathery haze at this stage. These long, feathery structures help in the wind dispersal of the seed enclosed in the base of each fruit. The Apache plume develops similar fruits, but the hairs on the styles are conspicuously longer and softer. In addition, the Apache plume has many more than ten pistils in each flower.

Cliffrose

Within the Great Basin, the cliffrose reaches its northernmost extension in the mountains of the central and eastern portions. It ranges from the lower mountain slopes and rocky canyons up to elevations of 8,000 feet within coniferous forests. Outside of our area, to the south, it is very abundant, becoming one of the dominant shrubs in the Grand Canyon. To the east it reaches New Mexico and Colorado.

Another interesting form, known as variety *dubia*, occurs only in the Providence Mountains of southern California. Per Axel Rydberg, a student of western botany early in this century, suggested that this form, characterized by much shorter styles, was actually a hybrid with the bitterbrush.

There are only two other species of cliffrose in the United States. One of these is a relict form known only from one locale in Arizona, while the other occurs in Texas. As a browse plant, the cliffrose has a good reputation, except for its brittle nature. It is considered highly desirable by deer and important for livestock. Like the bitterbrush and Apache plume, the cliffrose is a member of the rose family.

The cliffrose is a shrub well worthy of cultivation. Apparently, a few nurseries in California and Nevada regularly stock it. Although one has to go as far south as Mina, Nevada, to find it in the western part of the Great Basin, it will grow in the Reno area. There is, in fact, a herbarium record, albeit somewhat dubious, of its occurrence 8 kilometers north of Reno!

The common name of cliffrose comes from this shrub's preference for rocky canyon walls and the roselike appearance of its flowers. Another, older name was quinine bush, derived from the bitter taste of the twigs. The genus name *Cowania* commemorates James Cowan, a British merchant and amateur botanist, who introduced a number of Peruvian and Mexican species into England. The species name *mexicana* denotes the major range of the species proper in Mexico. The variety name *stansburiana* comes from the collector Stansbury, who provided the material, the so-called type specimen, that John Torrey drew upon for his description. The type locality is now called Stansbury Island, in the Great Salt Lake.

Ocean Spray
Holodiscus dumosus

THE CREAM-WHITE CLUSTERS of flowers borne on curved branches account for the common name of this shrub. Two other frequent names are creambush and rock-spiraea. The latter name is particularly appropriate for our Great Basin examples, since it connotes the usual rocky ledge and cliff habitat of this species. Spiraea is also a logical name, since ocean spray closely resembles that common cultivated shrub, except for its flower color, and it is, in fact, a member of the same subsection of the rose family. However, because of its specific habitat requirements, this is not a shrub which the casual traveler will be likely to see from the roadside.

Its individual flowers are quite small, possessing petals only about 2 millimeters long—however, the hundreds of flowers in each inflorescence make the ocean spray quite conspicuous. Blooming begins in June at the lower elevations but may not occur until August at the upper limit of its range. The petals and sepals are borne at the edge of a disk, and this particular structure accounts for the genus name *Holodiscus*, which comes from the Greek and means whole disk; the species name *dumosus* means shrubby. The fruits are small, dry, one-seeded structures.

Ocean spray ranges from .3 to sometimes over 1.5 meters tall. It is intricately branched and bears small, wedge-shaped leaves toothed at the top—under a hand lens the leaves can be seen to be pubescent on both surfaces. The smaller branches are reddish to tan in color and also somewhat pubescent. The Great Basin form occurs from 4,500 to 11,000 feet and is found consistently in rocky habitats. Another species, *H. discolor*, creambush, which occurs near the coast west of the Cascade Mountains, occasionally reaches a height of over 6 meters. Those species of ocean spray found in the Sierra Nevada and the Coast ranges occupy a quite different habitat, in

Ocean Spray

places growing as a common understory shrub in ponderosa pine forests. The genus, restricted to western North America, consists of two to five species, depending on which authority one follows.

Our species, *dumosus*, is quite variable, and some manuals recognize a separate species called *microphyllus* as an inhabitant of the Sierra Nevada and the Great Basin. However, although no detailed studies have been carried out on ocean spray, it seems most probable that the variations which have been described by some botanists are either the result of some environmental variable or constitute heritable variations below the species level. If these heritable variants occupy essentially the same kind of habitat, we refer to them as biotypes. If, on the other hand, they occupy obviously different habitats, we call them ecotypes.

Variation of the kind which occurs in ocean spray is especially characteristic of those species which occupy such specialized niches that large populations in any individual location are the exception. Inevitably, in this kind of situation, a phenomenon known as gene drift occurs. Essentially, this involves the random loss or addition by mutation of a number of genes or hereditary units in each population. The final form of any plant is ultimately the result of interaction between a set of genes and the environment. Consequently, the smaller the population and the more isolated it is from neighboring populations of the same species, the more probable it will be that, after a few generations, enough gene drift will have occurred to produce a recognizably different form. Given enough time, this evolutionary process can result in a new species, although whether it has yet done so in the case of ocean spray appears to be a matter of opinion.

Another interesting aspect of this kind of evolution is that natural selection plays little role in the production of the apparent variations by means of which we recognize these different forms. Natural selection is commonly described as the "sieve" which separates those heritable changes that, by chance, better adapt the organism to its habitat. In plants, at least, many of the characteristics used to separate one species from another have no apparent adaptive value, and are probably the result of random mutations or gene drift or some more complicated, but still largely chance-controlled hereditary mechanism. This is not to say that natural selection plays no essential role in the evolution of plants. As was pointed out in the introduction to this book, there are numerous morphological and physiological features which have been shown to be adaptive, such as leaf pubescence, succulence, or ability to withstand high salt concentrations. But, every plant has at least some features which no experimentation has yet shown to be of specific adaptive significance to a particular species. This would include such things as the size or number of serrations along the edge of a leaf, color of the epidermis on young twigs, or the value of an elliptically shaped leaf as opposed to one with a lancelike shape. In the latter instance, for example, both types of leaves would function to carry out photosynthesis, but a slight difference in shape probably has no critical adaptive value, even though it may well constitute one of several specific differences between two forms. This attitude toward the function, or lack thereof, of certain features in plants, is really a reflection of the modern scientific view concerning the major role played by chance in all natural phenomena. This is a nearly complete reversal of the philosophical attitude prevailing during the nineteenth century, when every structure in plants from the macroscopic to the microscopic was assumed to have a purpose, and a major activity by botanists was the deduction of these purposes. But, as is so often the case in nature, because an explanation seems logical does not imply that it is the correct or only explanation. Only experimentation can resolve the issue. Witness this quotation from Kerner von Marilaun and Oliver's two-volume work *The Natural History of Plants* published in 1895: "For us no fact is without significance. Our curiosity extends to the shape, size, and direction of the roots; to the configuration, venation, and insertion of the leaves; to the structure and colour of the flowers; and to the form of the fruit and seeds; and we assume that even each thorn, prickle, or hair has a definite function to fulfill." Random events or chance thus played no role in nature. Everything was basically a machine and imagined to be very efficient if not perfect in its assigned role. This extreme position was well expressed by one scientist who opined that "there is no essential difference between describing the trajectory in which a projectile moves on the one hand, and describing a beetle or the leaf of a tree on the other."

Dwarf Ninebark
Physocarpus alternans

THE WESTERN DWARF NINEBARK, like its eastern counterpart, is easily recognized by its bark—which tends to peel off in narrow strips, exposing the lighter bark beneath. Within the Great Basin it is most easily confused with the wax currant, which has leaves of similar size and shape. However, the latter does not have bark which peels in the fashion of ninebark, and, of course, the fruits of the wax currant are easily distinguished from the dry capsules of the ninebark. Also, the young branches and leaves of the ninebark are pubescent, with star-shaped or stellate hairs seen under a hand lens, totally unlike the simple pubescence on the wax currant.

Dwarf ninebark is a low shrub ranging from 25 centimeters to a little over 1 meter high in favorable locations. The rounded leaves vary from 6 to 20 millimeters long. Frequently, they are three-lobed. The clusters of three to six white flowers are borne at the ends of the smaller branchlets. The petals, only 4 to 5 millimeters long, are borne at the edge of a small cup, in a fashion that is similar to those of many other members of the rose family. The fruit which develops from the single pod is an inflated capsule about 5 millimeters long. The genus name *Physocarpus* comes from the Greek *physa*, meaning bellows, and *karpos*, meaning fruit. The pods or capsules are hygroscopic—splitting open at the tip during dry weather to release the seeds but closing during wet weather. This kind of response is common in many plants, all the way from mosses to higher plants. Dispersal of seeds or spores during dry weather generally means that they will travel farther and thus be more likely to colonize areas with similar habitats, and this confers an obvious evolutionary advantage on such a population. The species name *alternans* is in reference to the twenty or so stamens, some of which are longer with supporting stalks or "filaments" somewhat wider at the base.

The common name of ninebark is assumed to have originated from the numerous layers of bark which successively peel from the shrub's stems, although one authority believes that it originated from the number of medicinal uses for which the plant was noted. However, the western ninebark seems to have been little used—the Indian Medicine Project of the U.S. Department of Agriculture during the 1930s turned up no evidence that it was employed by Great Basin tribes, probably because it is not a common plant anywhere in the area.

The western ninebark inhabits dry, rocky slopes within and above the pinyon-juniper zone to an elevation of 10,000 feet. Altogether there are about fourteen species of ninebark, thirteen of these in North America and one in eastern Asia, according to many manuals. However, John and Rosemarie Kartesz, in a recent book, *A Synonymized Checklist of the Vascular Flora of the United States, Canada, and Greenland*, reduce this number to five, since only minor differences separate most of the so-called species. To the amateur, this kind of situation makes it appear that botanists are pretty contentious or at least that they have trouble getting their act together! Actually, it indicates that the concept of a species is a human invention—and not necessarily one that fits the natural situation in all instances. We like to pigeonhole things in order to think about them more easily, but more often than not in nature a continuum is the case. It should not be inferred from these remarks that the concept of a species is not a worthwhile artifice, for without it we would be at a loss to understand nature, but it is essential to remember that it is an abstract notion, very much like our notion of the electron!

The one species of ninebark which occurs in eastern Asia is an example of a disjunct distribution and an indication that, before the last series of glaciations, the ninebarks were more numerous and more widely distributed over the world. This is yet another example of the close affinity between the eastern temperate Asian flora and the flora of eastern North America. Tulip trees, magnolias, witch hazels, and a number of other genera show a similar disjunct distribution between eastern North America and eastern Asia.

Desert Peach
Prunus andersonii

Another shrub among sagebrush, antelope brush, and rabbit brush, particularly near Reno, is the wild peach, whose pink flowers make the roadsides flame. —Anonymous, NEVADA, A GUIDE TO THE SILVER STATE

WHILE IT IS somewhat hyperbolic to characterize our desert peach as capable of producing flaming roadsides, it is, without question, one of the most beautiful and probably the most underappreciated shrubs in the western Great Basin. It has also been called the desert wild almond, although one rarely hears that name now. The deep to light rose-colored or, rarely, white flowers are only 12 to 22 millimeters in diameter, but they occur in such masses that the shrubs for one or two weeks in any locale appear to be covered with a pink carpet.

Typically, the desert peach is a rigid, intricately branched shrub a meter or less in height, although it may on occasion become 2 meters high. The leaves, which appear in April or May at the same time as the flowers, are narrow and pointed and up to 2.5 centimeters long, being grouped in clusters on short, lateral branchlets. Each of the smaller branchlets ends in a spine, giving rise to the characteristic thorny aspect.

The fruit, which develops within a few weeks after flowering, resembles a small, fuzzy peach between 10 and 18 millimeters long, but unfortunately only a thin, inedible pulp surrounds the pit. Apparently the Paiutes made some use of the plant itself, for the leaves and twigs were boiled to prepare a tea considered efficacious in the treatment of colds and rheumatism. Our closely related western chokecherry was used by them in the same manner.

The desert peach, although common in the extreme western Great Basin, has a relatively restricted distribution, confined largely to the eastern slopes of the Sierra Nevada in California and Nevada. Its range extends from extreme northeastern California to Kern County in the south, then east through western Nevada to Churchill County. "Desert" peach is a misnomer, since this shrub is more at home in the steppe climate favored by the sagebrush. It is particularly common in the Reno–Carson City area. Even though it may be found on relatively dry slopes, it is not really capable of withstanding the very arid and saline environment of the desert.

B. L. Kay, J. A. Young, C. M. Ross, and W. L. Graves have studied the growth requirements of the desert peach in some detail. Apparently it does better and is most common on soils derived from decomposing granite. They found that it generally occurs as large clones, with individual plants being connected by underground stems—a single clone may cover several acres. Kay and his coworkers counted the annual rings in the stems of large clones and found none older than eight years. The clones themselves must be very much older, of course, but individual stems apparently are relatively short-lived.

Frank Vasek recently studied clones of the creosote bush at several sites in the Mohave Desert of California and found that the largest clone, with a diameter of 15 meters, approached an age of 11,700 years! Individual seedling stems of creosote bush varied in age from 19 to about 60 years. The larger stems split radially into segments which eventually gave rise to separate plants and to the resultant clones. Obviously, the bristlecone pine is not the oldest living thing after all! This is not to say, of course, that the clones of the desert peach are as old as those of the creosote bush, but who knows? Interestingly enough, individual clones of the desert peach differ considerably in flower color and abundance, in time of flowering, and undoubtedly in other ways. The variability in flowering time probably insures that at least some fruits will develop if a late spring frost occurs. Kay and his colleagues reported that individual clones in the same area might differ in flowering time by as much as a month.

Kay and his coworkers also looked at seed germination in the desert peach and found that, in order for the seeds to grow, they had to be exposed to a low temperature for several weeks. The best germination figure—44 percent—occurred when the seeds were kept at 2 degrees C. for four weeks. This process of exposing seeds to cool, moist conditions (typically in soil or

Desert Peach

in some medium such as charcoal or vermiculite) in the presence of oxygen is called stratification. Many temperate-zone plants produce seeds which must have their dormancy broken in this fashion. Obviously, this is an adaptive device which insures that the seeds will not begin to grow as soon as they are mature but, rather, will wait until the following spring. In some way not yet understood, exposure to cold in seeds such as this turns on certain genes responsible for the production of growth regulators which are necessary for the embryo to begin its development.

The genus name *Prunus* is the old Latin name for plum. There are about 150 species in the genus, distributed mainly in the northern hemisphere and including a large number of economically valuable forms of peaches, cherries, plums, apricots, and almonds. The species was named *andersonii* by Asa Gray to commemorate an early Nevada botanist, Charles Lewis Anderson, who practiced medicine in Carson City from 1862 to 1867 and collected intensively, published a "Catalog of Nevada Flora" in 1871 which listed those plants he had found near Carson City. The specimen from which Gray drew up his original description came from near Carson City.

Desert peach belongs to the rose family, the Rosaceae, which contains well over three thousand species distributed throughout the world. This family is especially abundant in the moister areas of the north temperate zone. While we have other shrub members of this family in the Great Basin, none of them is really adapted to the driest environments—though outside of our area, in southern Nevada, the blackbrush, which also belongs to the rose family, certainly comes close to being a true desert shrub. Like most members of this family, the desert peach is insect-pollinated, but its broad, open flowers permit any one of a variety of pollinators to receive a reward. There appears not to be the very specific adaptation to one or a few species of insects that we find in flowers with a closed aspect and accessory features such as spurs.

Bitter Cherry
Prunus emarginata

ANOTHER WESTERN mountain shrub which just barely makes it into the Great Basin is the bitter cherry, found on the eastern slopes of the Sierra Nevada in several of the western counties of Nevada. Bitter cherry is an appropriate name, since all parts of this plant are extremely bitter—stems, leaves, and fruits. Skirting the Great Basin, it extends north into British Columbia and east through Idaho to Montana. Around the southern end of the Basin, it gets into the cooler parts of Arizona. At its southern limit in California, bitter cherry may be found as high as 9,000 feet, while in British Columbia it ranges down to sea level. Its common associates in our area are ponderosa and Jeffrey pines, manzanita, tobacco brush, and serviceberry. Frequently abundant in very dense, uniform stands on steep, rocky slopes, it can also be found along valley bottoms next to streams, though usually not in any abundance at the latter locations.

Occasionally getting as high as 6 meters, bitter cherry is more typically a shrub 1 to 3 meters tall. The deciduous leaves, blunt and oblong, are up to 5 centimeters long and fine-toothed along the margin. Along with those of the trembling aspen, bitter cherry leaves turn yellow in the fall. Particularly conspicuous at that time are the bright red to black berries, borne singly or in small clusters. The twigs have a thin, reddish bark which when scratched produces an obvious cherry odor.

The clusters of white flowers, each a centimeter in diameter, are produced in the spring after the leaves have appeared. Each flower, in keeping with the pattern for the genus *Prunus*, has five petals and numerous stamens. The single pistil in the center of the flower eventually matures to form the fruit, which is like that of a typical cherry or peach. That is, the outer fleshy part surrounds the hard, stony pit that encloses the single seed.

Bitter Cherry

The only other shrub likely to be confused with the bitter cherry in our area is the sierra coffeeberry. They both have similarly shaped and sized leaves, but bitter cherry has one or two small, yellowish glands near the junction of the leaf blade and leaf stalk. Coffeeberry lacks any such glands. Also, the leaves in the coffeeberry are obviously thicker, with very prominent parallel, lateral veins running out to the edge. The flowers in the coffeeberry are small, green, inconspicuous affairs, and, while the fruits are similar in size and appearance, the berries in the coffeeberry contain two little nutlets rather than a single pit.

Despite its bitter nature, this shrub is browsed extensively by sheep, cattle, and deer. Apparently sheep are sometimes poisoned by it during the fall. However, there is some doubt about this, since bitter cherry has sometimes been confused with the chokecherries, which are definitely poisonous at certain growth stages, due to the hydrocyanic acid which seems to form in crushed or wilted foliage.

The genus name *Prunus* is the Latin term for plum, while the species name *emarginata* refers to the sometimes notched tips of the leaves.

Bitterbrush
Purshia tridentata

ALMOST AS apparent a part of the Great Basin as sagebrush is bitterbrush or antelope bitterbrush. By the pragmatically inclined, however, it is revered a great deal more than sagebrush because of its palatability for livestock and wildlife. Even the seed forms a major part of the fare of small animals. James Young and Raymond Evans, of the Agricultural Research Service at the University of Nevada, have pointed out that rodents and ants collect virtually the entire bitterbrush seed crop when it falls to the ground.

The natural range of bitterbrush extends from western Montana and British Columbia south to the arid portions of California and New Mexico, including the Rocky Mountains and, of course, the Great Basin. Although good estimates of its present extent in the eleven western states are unavailable, bitterbrush undoubtedly still covers tens of millions of acres and remains one of our most ubiquitous shrubs, next to the big sagebrush. Bitterbrush can be found in an enormous variety of habitats—ranging from arid flats, provided they are not saline or as dry as the shadscale association, to alpine zones well above timberline. In favorable locations, it will get to be nearly 3 meters tall, while above timberline it hugs the ground and on windswept slopes may be only 15 centimeters high.

Dwight Billings, who published a number of studies on the plant ecology of the Great Basin, considered the bitterbrush to be a codominant, along with rabbitbrush, littleleaf horsebrush, and green ephedra, in what he termed the sagebrush-grass zone. As the name implies, grasses are also an important element in this zone, especially various species of wild rye, wheatgrass, squirreltail, bluegrass, and needlegrass. This zone, which occurs above the shadscale zone, occupies more area in the Great Basin than any other vegetation zone.

Bitterbrush

Much of what we know concerning the ecology of bitterbrush is the result of the studies of Eamor C. Nord during the 1950s, when he was on the staff of the Pacific Southwest Forest and Range Experiment Station. Among other things, Nord attempted to study the shrub's ecological life history and the effects of various environmental factors on such things as its growth, development, longevity, and relationship to other plants. He picked twenty-nine locations in a variety of habitats in California for analysis.

One question which Nord was able to answer involved whether or not the various forms of bitterbrush are genetically determined—that is, are they ecotypes? He grew seedlings from erect, intermediate, and semiprostrate plants under uniform environmental conditions. Interestingly enough, seedlings from semiprostrate parents developed into erect plants 26 percent of the time, but erect parents produced only erect progeny. From this it appears that form in the bitterbrush is to a major extent determined by heredity. Nord found that the squat plants above 8,000 feet in the White Mountains averaged 52 years old, were over 2 meters across at the crown, and were less than 30 centimeters high. One plant was 115 years old and about 25 centimeters tall but covered an area of only about .6 square meter, while one which was 57 years old covered 12.5 square meters.

Bitterbrush roots penetrate to greater depths than those of sagebrush. Nord found roots 1.3 centimeters in diameter at a depth of 5 meters, indicating that they probably penetrated much deeper. This characteristic would allow bitterbrush to absorb water from deeper zones within the soil, even though adjacent sagebrush plants may show water stress as a result of their shallower root system.

Bitterbrush begins to leaf out in the spring about two months sooner in the southern part of its range in California than in the northern part of the state. In nature, plants are not mature enough to produce flowers and seeds until they are about ten years of age. At low elevations and to the south, seeds mature in June, while at the opposite extreme in elevation and latitude they may not mature until August or September. Once mature, the seeds drop from the plant in a very few days. On favorable sites, seed production may be as high as 500 pounds per acre. Abundant seed production appears to occur the year after large amounts of precipitation. The seeds are heavy and cannot be carried far by the wind; apparently only rodents and birds carry them any distance—and then usually not very far.

Actually, these seeds comprise the entire dried pistil with a seed enclosed, a so-called achene. The dried parts of the flower usually remain attached when the achene falls. A. L. Hormay found that these dried flower parts contain a strong germination inhibitor. Janice Beatley discovered that desert rodents carefully remove the flower parts when they cache the seeds, at a depth of about 5 centimeters, in the soil; the rodents return to these caches in the spring, when the seeds germinate, to consume the cotyledons. Apparently, the seedlings are a good source of the carotene needed for the rodents' reproductive cycle. James Young and Raymond Evans studied the germination requirements of bitterbrush and found that cool, moist conditions were needed for some weeks before the seeds would grow. The temperature had to remain between 2 and 5 degrees C. Once the seeds were exposed to these conditions for four weeks, they were able to germinate under a wide range of temperatures. Obviously, this duplicates the conditions to which the seeds would be exposed over winter. Germination inhibitors, very common in temperate-zone plants, serve to insure that growth will not begin during the wrong season.

Despite the large amount of achenes produced, Nord found that the number of seedlings appearing in the spring varied enormously from site to site. The number of seedlings dropped abruptly when the total annual precipitation was below 18 centimeters, with little or no snowpack. Some sites had no seedlings at all, while one site in Mono County had more than ten thousand per acre. Nord discovered that seedlings rarely occurred near older or larger bitterbrush plants; usually sagebrush was the nurse shrub. He suggested that the litter under bitterbrush may contain an inhibitor which prevents the growth of its own seedlings. Once established, the roots of the seedlings grow very rapidly, attaining a length of half a meter the first season.

In addition to seeds, bitterbrush is able to reproduce effectively by means of stem layering. This results when those lower branches which chance to be in contact with the soil produce adventitious roots. Eventually, the connection with the parent plant may be severed. Layering is most frequent at high elevations and on some burnt-over areas; at 10,600 feet on the White Mountain site, 80 percent of the bitterbrush had stem layers. At lower elevations, the figure was 40 percent or less.

Following fires, bitterbrush may resprout from the roots, depending on the amount of soil moisture. Nord points out that a number of observations attest that ample soil moisture improves the ability of bitterbrush to resprout. A burn near Carson City, Nevada, for example, effectively destroyed the aerial parts of bitterbrush plants, but the shrubs nevertheless resprouted, apparently because the fire was put out by a heavy rainstorm. However, in many instances, bitterbrush will fail to make a significant recovery following fire. Bitterbrush on pumice soils appears to make a faster recovery than it does on other soil types. In some instances, it appeared to Nord that other vegetation which came in immediately after a fire helped bitterbrush become established—even though initially the former must have competed for the environmental essentials.

Bitterbrush is a pioneer shrub at seemingly hostile sites, although it will not grow on calcareous or saline soils. It was one of the first perennial species to become established after the eruption of Mount Lassen from 1914 to 1916. Ultimately, bitterbrush helps the succession process for other plants by assisting in soil retention and development and by adding organic debris to the soil. Nord studied the vegetation of Panum Crater in the Mono Crater formation to learn something about the role played by bitterbrush in succession. Apparently, this crater was formed during the late Pleistocene. The upper rim of the crater, which is composed of deep pumice fragments, is dominated by bitterbrush—it made up about 70 percent of the stand, with associated desert peach and rabbitbrush making up the remainder. Nord found the most ancient bitterbrush known, at least 162 years old, at this site. Even more remarkable for this particular site were the vigor and persistence of the bitterbrush. The area has been grazed for the past century, yet desert peach and rabbitbrush, which are unpalatable, had not made significant incursions into the bitterbrush. Heavy grazing, of course, will be detrimental. Sheep, especially, tend to eat the young plants and buds. Cattle, on the other hand, graze on the more mature plants. Nord found that moderate grazing appeared to encourage the production of seedlings.

B. R. McConnell and G. A. Garrison, studying the seasonal variation of available carbohydrates in bitterbrush, found that beginning in the spring there was a decline in carbohydrates which continued through the growing season until mid August. From then until leaf fall, the carbohydrate reserves continued to build up. Obviously, this research indicates that it would be best not to use bitterbrush as a browse in midsummer. Aside from carbohydrates, bitterbrush is a good source of crude protein and fat.

One very significant aspect of shrub adaptation to the environment is the presence of mycorrhizae—small roots which are infected with fungi. The fungus may be prolific enough to form a sheath around the root. Far from being an infection which harms the plants, it is in many cases an absolute requisite for survival. The fungi take the place of root hairs by absorbing water and nutrients from the soil and supplying these essentials to the host plant. In addition, recent research has shown that the fungus also supplies some growth hormones to the host. It is thought that mycorrhizae increase chlorophyll content as well as drought and temperature resistance; they may even increase resistance to some disease organisms. The fungus in turn gets its basic nourishment from the host. This kind of symbiotic relationship, from which both partners benefit, is very common in nature.

In the case of bitterbrush, Stephen Williams, working with plants collected in Wyoming, found that the roots were heavily infected with mycorrhizae. Not only do the mycorrhizae confer the benefits already mentioned on the bitterbrush host, but Williams thinks it probable that the mycorrhizae in this case are responsible for fixing atmospheric nitrogen and making it available to the host in a fashion similar to that carried out by root nodule bacteria in the pea family. However, no evidence is yet available to indicate how much nitrogen may be contributed to the soil by bitterbrush.

Bitterbrush typically varies between 1 and 2 meters tall, although in favorable locations heights of nearly 3 meters may be attained. Its leaves are small, wedge-shaped, and three-lobed, between 1 to 2.5 centimeters long. They are bright or dark green and, under a hand lens, can be seen to be finely pubescent on the upper side. Beneath they have a short, white pubescence, and the margins are slightly curled under. Most of the leaves are lost with the onset of winter.

The flowers, 1.5 to 2 centimeters in diameter with pale yellow petals, are borne on the stems at the base of the leaf clusters. Depending on the elevation, blooming occurs from early spring to July. The fruit—a small, one-seeded, leathery structure 8 to 12 millimeters long—falls entire from the plant without releasing the enclosed seed.

The large number of studies on bitterbrush attest to its importance as a native browse plant for wildlife and livestock. H. Bissell and his coworkers found that deer could exist for some time feeding only on bitterbrush. The proportion of crude protein in the leaves ranges as high as 14 percent in early summer. Antelope, elk, bighorn sheep, and moose utilize bitterbrush extensively, and the seed forms a significant portion of the diet of rodents, ants, and birds.

Great Basin Indians made extensive use of bitterbrush as a remedy for chicken pox, smallpox, measles, and venereal disease. A solution prepared from the leaves was thought to be a good antiseptic for rashes, scratches, and insect bites. Tea prepared from the leaves or bark was believed to be useful against tuberculosis, pneumonia, and colds.

The only other species in the genus is the desert bitterbrush, *P. glandulosa*, sometimes called the Mohave antelope brush. This species occurs in the arid portions of southern Nevada and California at the edge of the driest part of the desert and near the chaparral zones at higher elevations. Desert bitterbrush can be easily distinguished from the Great Basin species by the former's narrower, more deeply lobed, and glandular-punctate leaves. Unlike the antelope bitterbrush, desert bitterbrush is evergreen. Where the two species overlap, they are known to hybridize and intergrade. On this same note, hybrids have been described from a number of locations where bitterbrush and cliffrose, *Cowania mexicana*, overlap. Hybrids between species of many plants are quite common, but they are somewhat less so between genera. Irving Knobloch of Michigan State University has recorded some 2,993 reports of intergeneric hybridization, however. In fact, G. Ledyard Stebbins presents good evidence to indicate that the desert bitterbrush is actually a stabilized hybrid between the antelope bitterbrush and the cliffrose which may have evolved at the end of the Pleistocene wet period in the Great Basin.

Howard Stutz and Kay Thomas of Brigham Young University conducted extensive studies on populations of bitterbrush and cliffrose throughout Utah, Idaho, Montana, and California. They found that hybridization between these two species was common wherever they grew together. When this sort of hybridization occurs in nature, a frequent consequence is the backcrossing of these hybrids to either parent species, with the result that after a few generations many of the characters of one species begin to be apparent in much of the population of the other species. This phenomenon of gene flow from one species to another by means of hybridization is known as introgression. Stutz and Thomas found that introgression of this sort was so common that nonintrogressed populations of bitterbrush were rare or absent in Utah. Although the apparent barrier to crossing in these two species was a difference in their flowering periods, rough terrain with differing exposures permitted an overlap in the flowering times, with the subsequent production of hybrids. Even in Idaho and Montana, far north of the cliffrose range, it was possible to find cliffrose characters in bitterbrush populations. The authors postulate that natural selection favored this widespread introgression, possibly because of the unpalatability of cliffrose. This may account for the unpalatable bitterbrush populations reported throughout Idaho, Washington, and Montana.

Based on the structure of its flowers, bitterbrush would be placed among the less evolved members of the rose family. The genus name honors Frederick Traugott Pursh, an early American botanist who published *Flora Americae Septentrionalis* in 1814. He was the author of the western genus *Lewisia*, known popularly as the bitterroot, named in recognition of the explorer Meriwether Lewis. The species name *tridentata* is the Latin term for three-toothed, in reference to the three-lobed leaves. A plethora of other common names have been given to bitterbrush, including deerbrush, buckbrush, and quininebrush. Some have even called it by the inappropriate names of greasewood and black sage!

Wild Rose
Rosa woodsii

The Rose doth deserve the chiefest and most principall place among all floures whatsoever; being not only esteemed for his beautie, vertues, and his fragrant smell, but also because it is the honour and ornament of our English Sceptre. —Gerard's HERBALL as quoted in Alice Morse Earle, SUN-DIALS AND ROSES OF YESTERDAY

WHILE IT IS found throughout the Great Basin, the wild rose is hardly a plant of desert habitats, preferring rather moist situations along streams or in seepage areas, along fences, and in other protected spots. Of all our shrubs, this is the easiest for anyone to recognize. Like most species of wild and cultivated roses, our form has the conventional divided leaves, thorny stems, and pink flowers about 40 millimeters or so in diameter; it differs from the familiar cultivated rose in having only five petals. When conditions are favorable, nearly impenetrable thickets of roses are formed along some of our mountain streams. Following the brief blooming period in the spring, the bright scarlet fruits or rose hips ripen during the late summer and fall.

Not uncommonly, rose hips will remain on the plant after the leaves have fallen. They are an important part of the diet for various birds and mammals. In reality, the hip is an urn-shaped structure, in the cavity of which are borne the true hard, woody fruits or achenes. The urn, which becomes fleshy at maturity, has been used to prepare preserves, tea, and candy. The leaves were used by Great Basin Indians to prepare a general tonic. They also regarded portions of the bark, roots, and stems as a very beneficial dressing for wounds, sores, and burns, and the roots, especially, were seen as useful in curing diarrhea. Rose hips do contain vitamin C, and possibly this accounts for some of the apparent benefits. Europeans, similarly, developed

Wild Rose

a great variety of uses for the rose and acknowledged its beauty in much of their literature. It was once the practice to suspend a rose from the ceiling when conversations were to be kept secret, accounting for the term sub rosa. This custom is attributed to the legend that Cupid bribed the god of silence with a rose.

The genus *Rosa*, which occurs on all the continents of the north temperate zone, contains well over 250 species. Most of our cultivated roses were derived by selection and hybridization from Asiatic forms. The birthplace of the rose was probably Iran, from which it was carried gradually to southern Europe, to become a favorite flower of the Greeks and Romans. This selection process involved the development of the first known double flowers, in which the bloom belies the simple form of its progenitors by its multiplicity of petals. In the rose, it is thought that many of the inner petals are really stamens which have been modified into petals by some unknown genetic process. This is not so strange as it might seem, for we now believe that primitive flowers had many petals and many stamens and that the outermost stamens were essentially petallike and had only poorly developed anthers. In a real sense, in cultivated roses, we have caused a reversion to a more primitive form, albeit one that does not now occur in the rose family but is, instead, typical of the magnolia family.

FABACEAE
PEA FAMILY

Smokebush
Psorothamnus polydenius

ONE OF THE MOST fragrant of our desert shrubs, with an odor which resembles that of some exotic member of the citrus family, is the smokebush. Known in older literature as *Dalea polyadenia*, the smokebush is most frequent in the Carson Desert region of the Great Basin. Its characteristic fragrance is the result of the secretion of volatile substances by the numerous yellow to orange, pinhead-sized glands scattered over the light green stems. Aside from these stem glands, which none of our other shrubs possesses, the smokebush is easy to recognize because of its light, angular, stiff stems, which are only sparsely leafy, even during the wettest season in the desert. It may attain a height of between .5 to 1.5 meters.

The smaller stems of the smokebush often taper to a spine-tipped end. The leaves, produced in the spring, consist of divided structures only 12 to 25 millimeters long and composed of tiny leaflets. By midsummer the leaves have generally dried and fallen from the plant. This summertime loss of leaves is a common water-conserving procedure of many desert shrubs. However, an unusual wet period during the summer will result in a second crop of leaves. Even when no leaves are present, considerable photosynthesis is probably carried on by the chlorophyll-containing stems.

We know a fair amount about the life history and ecology of the smokebush, unlike most of our desert shrubs, thanks to a very complete study carried out by Irwin Ting, who worked on this problem under the author's direction for his master's degree. The smokebush is most commonly found in sandy areas and, depending on the locale, has as its associates the four-winged saltbush, Shockley's desert thorn, rabbitbrush, Bailey's greasewood, or hairy horsebrush. Many plants, such as the smokebush, which prefer sandy habitats apparently do so because of the need to have their roots well

Smokebush, winter aspect

aerated. In order to measure how much less dense these sandy habitats really are, Irwin and I used a simple apparatus consisting of an open-ended cylinder, a manometer, and a tire pump. The cylinder was driven into the ground and air was pumped into it until the pressure stabilized. We found that sandy loam was a full seven times less permeable than sand, while the clayey soil in a greasewood community was an average of five times less permeable. In reality, this latter soil was the densest, and the lower than expected average was primarily the result of the numerous cracks which develop as the soil dries out. At any rate, oxygen can get to roots in soil only by a slow diffusion process, so the more permeable a soil, the better off most plants will be.

Soon after the leaves develop, 1-to-2.5-centimeters-long clusters of small, purple flowers begin to form. Blooming continues until the first severe frosts of fall. The flowers are followed by small, one- or two-seeded pods, about 5 millimeters long. These pods typically do not open when ripe but instead fall from the plant with the dead sepals still attached.

A lengthy series of experiments was designed to test a variety of factors which might conceivably control seed germination in the smokebush—knowing something about germination requirements frequently tells us why some plants may be found only in certain locales. Unlike many seeds, which must undergo a period of afterripening before they will begin to grow, smokebush seeds are capable of immediate growth, showing signs of growth as soon as twelve hours after being moistened. In many other temperate-zone plants, the seeds must be exposed to a few weeks or longer of cold weather before they will germinate. Although smokebush seeds could withstand extremes of cold below freezing and heat to 60 degrees C., exposure to such extremes did not hasten germination.

One factor which deterred germination was the seed coat, which had to be cracked or scratched in some way before the dormant embryos would absorb water and grow. We might expect that this scarification of smokebush seeds would be accomplished in the shrub's native habitat by blowing sand. Since the seeds are dispersed still enclosed in their tiny pods, Irwin thought that the latter might have something to do with germination. He discovered that the seeds would not germinate, even if scarified, if they were exposed to a red-orange extract which was obtained from the mature pods. Apparently two factors, then, prevent smokebush seeds from germinating too soon under natural conditions: the chemical inhibition of the surrounding pod and the physical restraint imposed by the hard, water-resistant seed coat. If the pods are kept moist, bacterial action and leaching will eventually destroy the chemical inhibition. This, of course, takes some time in nature—from a few weeks to months, depending on the weather. This kind of inhibition by the enclosing fruit is rather common; it insures that most seeds will not begin to grow while the fruit is still attached to the parent plant. Tomato seeds, for example, will germinate immediately, but only if all the surrounding pulp is removed.

Smokebush seeds twenty-three years old were still capable of germination, and, not too surprisingly, the impermeable nature of the seed coat apparently decreased with age. This could insure that plenty of seeds would be available for germination in favorable years. Apparently, a fair amount of moisture must be present in the soil before the seedlings can be established, though saturated sand cannot be tolerated, probably because the developing root system requires more oxygen than is needed by those desert shrubs growing in denser soils. In fact, much of the distribution pattern of the smokebush can be understood in terms of the requirements for germination and development of the seedlings. It really doesn't matter whether the mature shrub can tolerate a much broader range of conditions, for if the seedlings won't grow except in a restricted environment, then, of course, the mature plants will be restricted to that same environment.

Seedlings only one year old and about 3 centimeters tall had developed a taproot between 20 and 30 centimeters long. By two years of age, the root had grown to a depth of over 40 centimeters. However, lateral roots showed no significant development until after the second year. Apparently, most of the energy these young plants expend goes into the rapid development of a long taproot, which penetrates to the zone where water will be available to the seedling. While sand is a good medium for aeration of the roots, it has the disadvantage, so far as plants are concerned, of allowing water to drain rapidly to lower levels. Consequently, we find that many sand dune plants have roots penetrating to relatively great depths (although, we must be careful not to say that they have grown to such depths in their "search" for water!). One 60-centimeter bush had a taproot 162 centimeters long and side roots up to 330 centimeters long; its lateral roots were concentrated between depths of 30 to 60 centimeters. At least in these studies, that seemed to be the zone where most of the moisture was retained by the sand.

The age of smokebush plants of various sizes was determined by counting the annual rings of the main stem under a microscope. The longest-lived specimen was thirty-seven years old, while most of the mature plants in the sample tested were between twelve and twenty-five years of age. There is some uncertainty in these figures, since an unusually wet summer could lead to the formation of two rings in a single year.

The smokebush is a member of the pea family, the Fabaceae. This very large family, found all over the world, contains about seven hundred genera and seventeen thousand species. Although it is well represented in lower-latitude deserts, there are few shrubby members of the group in the Great Basin. The smoke tree, *P. spinosus*, of Arizona and southern California is a close relative. The genus consists of only nine species, confined to the Sonoran, Chihuahuan, and Mohave deserts in addition to the deserts of the Colorado Plateau and the Great Basin. It extends into northwestern Mexico, including Baja California. Rupert C. Barneby, of the New York Botanical Garden, is the world's expert on many genera of the pea family; he has recently completed an 891-page volume on *Psorothamnus* and its relatives called *Daleae Imagines*. Barneby considers *Psorothamnus* to be an ancient and primitive group, compared to certain other genera to which it is related.

The more advanced members of the pea family all have a basically similar flower structure, exemplified by the familiar sweet pea. The flowers have a bilateral symmetry; that is, each half is the mirror image of the other. The large upper petal in pea flowers is called the banner or standard, the two lateral petals are the wings, and the basal petal or keel really consists of two petals fused together which partially enclose the stamens and the single pistil. Characteristically, there are ten stamens; in many species nine of these are fused together at the base, with one stamen remaining free. Depending on the species, this simple bilateral pattern has evolved into a variety of mechanisms which insure that cross-pollination occurs. This kind of bilateral or zygomorphic flower is supremely well adapted for cross-pollination by various species of bees and butterflies.

The flower's lateral wings serve three basic functions—as a landing place for the insect, as levers to depress the keel and expose the receptive stigma of the pistil plus the pollen as a result of the weight of the insect, and as levers which help the stamens and pistil return to their original position after the insect visitor has left. The pollinator has to be the right size, shape, and weight or this elaborate pollinating machine will refuse to function. That it is only infrequently successful is evidenced by the relative rarity of seed production in this species. One of my colleagues tells me that, though the smokebush has not been well studied from this standpoint, the flowers appear to be visited by both generalists and specialists. The specialists would be those bees which visit only smokebush flowers and no others.

A reasonable question at this point might be, why do so many species go to such lengths to insure cross-pollination? Some plants appear to get along perfectly well with only self-pollination, and, in fact, seed production is frequently almost infallible in such populations. This would appear to confer a considerable advantage, since these plants are no longer dependent on insects for reproduction. But a price is paid, for such self-pollinating populations are less variable—in a sense, every individual is pretty much of a conformist. This is fine as long as things stay the same, but, of course, in nature this is never true. Sooner or later, the climate or soil changes, or a disease organism comes along, and our conformist population may suddenly become extinct.

On the other hand, cross-pollinated species tend to be much more variable, and mutations may spread much more rapidly through the population in the course of a few generations. There are many more eccentrics in such populations. Thus, when an environmental change does come along, at least some individuals may be able to make the transition. Given enough change, we may even wind up with a new species which has evolved in place of the old one. But, even though the old population has become extinct, at least it did give rise to another species rather than simply disappearing, as our conformist group did. Evolution, then, tends to favor cross-pollinated species, which explains why so many plants are that way. We have to be careful here not to imply that any population can anticipate environmental change. It is pure chance that determines whether or not any of the random eccentrics will have the right physiology or morphology to survive.

Possibly because of its odor, the various Indian tribes in the Great Basin made extensive use of the smokebush in attempting to cure disease. Commonly, it was used as a remedy for colds, a tea being prepared from the stems for this purpose. One of the most extensive studies on Indian medicine lore in the Great Basin was carried out by the Bureau of Plant Industry of the

U.S. Department of Agriculture, working with the Biology Department of the University of Nevada during the 1930s. Percy Train, along with James R. Henrichs and W. Andrew Archer, published the results of this research in 1941 under the auspices of the USDA. Agnes and Percy Train conducted most of the fieldwork with the Indians, recording information about approximately two hundred species considered to be of medicinal value. The Trains reported that various Indian tribes regarded the smokebush as efficacious in curing a number of ailments, aside from the common respiratory ones; these were most notably kidney problems, smallpox, venereal diseases, measles, muscular pains, and stomachaches. But, since it was not feasible to carry out any well-controlled medical studies, we really don't know whether smokebush was any good at all for these troubles. Agnes Train has written entertainingly of her collecting experiences with her husband in the silver state in a book entitled *Nevada through Rose-Colored Glasses*.

The genus name *Psorothamnus* comes from the Greek *psoraleos*, which means rough, and *thamnos* meaning bush. The species name *polydenius* means many glands, in reference to the oil glands on the stems. From the Humboldt and Carson sinks, the smokebush extends south to the Virgin and Meadow valleys. There is also a small population along the Green River in east central Utah. This latter population has been described by Barneby as a distinct variety, *jonesii*, named after Marcus Jones, a well-known botanist of the West who died in 1934.

One of the more spectacular smokebush species, with a distribution just south of the Great Basin in southern Nevada, Arizona, Utah, and California, is the Fremont dalea or smokebush, *P. fremontii*. This species produces conspicuous, elongated clusters of deep purple flowers from April to June. Another member of this genus, king's dalea, *P. kingii*, occurs on sand dunes in Churchill and Humboldt counties in western Nevada. It is relatively uncommon, except in the few specialized habitats where it is found, and at one time it was considered to be a threatened species. It differs considerably from the common smokebush, having a lower stature, only one to four flowers in each cluster compared to the eight or more flowers in dense terminal clusters in the common smokebush, and, most notably, reddish, cordlike stems located up to 12 centimeters below the sand surface. These subterranean stems or stolons allow this species to spread rapidly and to effectively combat the problem of drifting sand which any sand dune plant must face.

ELAEAGNACEAE
OLEASTER FAMILY

Silver Buffaloberry
Shepherdia argentea

Soups were popular, some brewed with fleshy ribs and assorted bones fractured to render up their marrow, others with buffalo meat, berries, fat, and the roots of jack-in-the-pulpits. —Tom McHugh, THE TIME OF THE BUFFALO

THE BERRIES referred to above were typically either serviceberries or the sour, reddish or golden yellow fruits of the buffaloberry—early settlers soon learned that the berries of both could be dried or made into a delectable jelly or jam. The silver buffaloberry is a shaggy-barked, thorny, deciduous shrub or small tree up to 6 meters tall. Both surfaces of the leaves are covered with microscopic, star-shaped scales that reflect the light and account for the shrub's rusty, silvery aspect. These scales undoubtedly help reduce water loss during the summer.

Within the Great Basin, silver buffaloberry is common along many streams, on moister hillsides, and frequently on valley bottoms where the soil is not too saline. It ascends to elevations of 7,500 feet on some of the interior mountain ranges. It is unusual among our berry-producing shrubs in that male and female flowers are borne on separate plants. Flowering occurs quite early in the season, sometimes as early as the end of February, before the leaves appear. Both male and female flowers are small, yellowish, and lacking in petals. The four sepals in the male flowers subtend the eight stamens. In the female flowers, the basal portion of the four sepals forms a cup which is fused with the lower part of the pistil—this basal cup forms the fleshy part of the fruit later on. The apparent seed within is really a kind of small nutlet that develops from the pistil proper.

Shepherds regard the buffaloberry as fair forage for sheep in the northern

Silver Buffaloberry

part of its range, but it is generally considered worthless for cattle. Deer and elk are said to make good use of it.

The buffaloberry is a member of the oleaster family, the Elaeagnaceae, which possesses about three genera and some fifty species. Many are thorny and, like the buffaloberry, have leaves covered with silvery or golden scales. Despite its small size, the family is widely distributed in the northern hemisphere, being found in North America, Europe, and Asia. In Australia, it occurs in the east and south. The family name comes from the Greek and means sacred olive, because of its members' resemblance to the true olive, to which, however, they are unrelated. A tree belonging to the family, *Elaeagnus angustifolia*, known commonly as the Russian olive, has become naturalized throughout the Great Basin; its original home is southern Europe and western Asia. Like the buffaloberry, its silvery leaves are quite distinctive.

There are two other species of *Shepherdia* in North America. One of these, *S. rotundifolia*, has leaves which are persistent (that is, tending to be evergreen) and, as the name implies, circular or oval in outline. It occurs from southern Utah into Arizona in the sagebrush and pinyon zones. The other species, *S. canadensis*, resembles the silver buffaloberry but lacks thorns and has leaves which are green rather than silvery on the upper surface. The Canadian buffaloberry occurs throughout Canada to Alaska and south to Washington, Oregon, Utah, and New Mexico, but it avoids most of the Great Basin proper. At its southern extremity, it is confined to the higher vegetation zones in the mountains. By comparison, the silver buffaloberry has a scattered distribution throughout the West from southern California throughout the Great Basin to Nebraska, the Dakotas, and south central Canada.

The buffaloberry's generic name honors John Shepherd, an English botanist, while the species name means silver in Latin. Other common names for the buffaloberry are Nebraska currant, rabbit berry, silverleaf, and wild oleaster.

CORNACEAE
DOGWOOD FAMILY

American Dogwood
Cornus sericea

ANOTHER COMMON NAME for the dogwood is red osier, and this is, perhaps, a better one since it connotes the very characteristic smooth, wandlike, red stems. These are not evident during the summer, when the shrub is clothed with its bright green leaves, but once the leaves have been lost during the fall the color and form of the dogwood appear particularly distinctive. Additional common names for the American dogwood are redbrush, gutter tree, and kinnikinic, though these are seldom used in our area. The latter name was also used by Indians to refer to a smoking mixture composed of tobacco and stem scrapings of the American dogwood.

No other native shrub in the Great Basin has such straight, brilliant red stems. The leaves are lance-shaped, without teeth along the edges, and borne two to each node on the stem. When two leaves are borne at each node, the condition is called opposite by botanists—opposite-leaved shrubs are relatively infrequent among Great Basin forms. To the unaided eye, the leaves appear hairless, but examination with a hand lens will reveal a fine pubescence on both surfaces. This pubescence may at times be fairly obvious; the more pubescent form, known as subspecies *occidentalis*, extends throughout California north to British Columbia. Hybrids between this form and the more typical eastern subspecies are common. The leaves also have prominent lateral veins which curve toward the apex at their extremities.

The American dogwood's small, white flowers are grouped together in flat-topped clusters 3 to 6 centimeters across. When viewed with a hand lens, the individual flowers can be seen to consist of four very small sepals, four white petals about 5 millimeters long, and four stamens all perched atop the ovary of the pistil. An arrangement of this sort, with the flower parts mounted on top of the ovary, is called inferior. This does not imply a value

American Dogwood

judgment—it merely indicates that the ovary is below the sepals and petals. There is good reason to believe that this is an advanced evolutionary condition among flowering plants.

In some members of the genus *Cornus*, the cluster of small flowers is subtended by bracts, really modified leaves, which fail to develop chlorophyll and instead expand to produce white, petallike structures when the true flowers open. The eastern flowering dogwood, *C. florida*, and its western counterpart the mountain dogwood, *C. nuttallii*, are particularly conspicuous in the early spring because of their large, white "flowers," each of which is really an entire cluster of small true flowers like those of the American dogwood. The flowers of the American dogwood are followed by white or bluish, berrylike fruits about 8 millimeters in diameter. Technically these are not berries, since the single hard seed inside is a pit, much like the pit of a cherry; the scientific name for this kind of fruit is drupe. Pits, of course, do have a seed inside them.

The American dogwood is a spreading shrub of streamsides and marshy areas from valley bottoms to mountain canyons at 9,000 feet. Frequently, dense thickets of American dogwood, alders, and willows will block the traveler's way along our mountain streams. It is a common shrub throughout the northern latitudes of North America, extending north to Alaska and east to Newfoundland and Virginia.

The family to which the American dogwood belongs is the Cornaceae. This is a small family, of about eleven genera and one hundred species, but with a very wide distribution in the temperate areas of all the continents except Australia. Because of this wide range, the family is regarded as probably being fairly ancient. *Cornus* fossils have been found in rocks of Eocene age, some 50 or 60 million years old. Arthur Cronquist of the New York Botanical Garden regards the Cornaceae as being derived from the rose order, with a common ancestor connecting it to the hydrangeas and currants. In fact, one genus of New Zealand shrubs known as *Corokia*, now placed in the dogwood family, was at one time considered to belong with the currants—Cronquist refers to it as a nonmissing link!

The genus name comes from the Latin *cornu*, which means horn. This is in reference to the wood of several species, which is sufficiently hard to have been used not only for furniture but also for durable weaving implements. The species name *sericea* is taken from the name for the fine hairs on the leaves and younger stems. At one time the American dogwood was known by another species name, *stolonifera*, which means stolon-bearing. Stolons are stems which run along or sometimes beneath the surface of the ground and which, by rooting at intervals, serve to vegetatively propagate a plant; the strawberry is a prime example of a stoloniferous plant. The American dogwood is also able to propagate itself by this means.

The common name of dogwood appears to have originated as a corruption of "dagger wood," since the wood was carved to make crude daggers or skewers. Another theory has it that in medieval times the bark of an English species was boiled and the extract used to treat mange in dogs, thus accounting for the common name. Our eastern flowering dogwood has been employed as an effective substitute for quinine. Various Indian tribes used an extract from the American dogwood and other species of dogwood as an emetic in treating fevers and coughs. They also made a black dye by boiling American dogwood bark with rusty iron, and they obtained a red dye by boiling the small roots. The cornelian cherry, *C. mas*, is a European species which produces scarlet berries long used for jams, preserves, and beverages. Because of its bitter taste, American dogwood is considered relatively unpalatable, even for sheep and goats. Sometimes the young sprouts, which are less bitter, are eaten by deer and cattle.

CELASTRACEAE
BITTERSWEET FAMILY

Spiny Greasebush
Forsellesia nevadensis

THE SPINY GREASEBUSH is almost never a very conspicuous or dominant shrub in the Great Basin. Generally, it occurs in relatively small numbers on steep hillsides or in ravines, mixed with other shrubs such as rabbitbrush, green ephedra, and bitterbrush. It appears to prefer limestone areas, though this is not always the case—one notable stand near Virginia City, Nevada, is located on soils derived only from igneous rocks. In its leafless condition, the spiny greasebush somewhat resembles a rather dark and stiffly angled green ephedra. However, closer inspection will show that many of the smaller branches end in spines, and this is never the case with green ephedra.

In the spring, the younger branches of the spiny greasebush produce small, oblong leaves from 5 millimeters to 1 centimeter long. The tiny, inconspicuous, white flowers, with five petals, are produced in April or May and are followed by capsules only 2 to 5 millimeters long, each bearing one or two seeds. During most of the year, leaves are not present; photosynthesis is carried on by the green, angular branches. Characteristically, spiny greasebush is only 30 to 60 centimeters tall. If it has any value as a forage plant, this has not been noted.

Despite its lack of abundance in any one area, the spiny greasebush has a wide distribution, ranging from southwestern California north to the drier eastern portions of Oregon and Washington and east through Nevada to Arizona, Utah, New Mexico, Oklahoma, and Texas. The Great Basin spiny greasebush was first described by the famous nineteenth-century American botanist Asa Gray, from material collected in northern Washoe County, Nevada. He described it under another genus name, however, calling it *Glossopetalon nevadense*.

The genus name honors James Henry Forselles, a Swedish mining engi-

Spiny Greasebush

neer and botanist of the nineteenth century. The alternate name, *Glossopetalon*, comes from the Greek and means tongue and petal, in reference to the shape of the petals. Lyman Benson and Robert Darrow, in their recent book, *The Trees and Shrubs of the Southwestern Deserts*, consider the Great Basin spiny greasebush to be only a variety of *Glossopetalon spinescens*; they use the variety name *aridum*, which refers to this shrub's dry habitat.

The genus *Forsellesia*, which contains eight species, ranges from Washington south to eastern California, east to Idaho, Nevada, Utah, and Colorado, and southeast through Arizona and New Mexico to Oklahoma and Texas. Asa Gray thought that greasebush belonged to the bittersweet family, known botanically as the Celastraceae. This family—widespread in the temperate and tropical areas of the world—includes such plants as the burning bush or euonymus, some exotic species of which are commonly cultivated, as well as Oregon boxwood, Utah mortonia, and the bittersweet of the eastern states. One Middle Eastern tree belonging to this family is the source of Arabian tea, though not the tea of the Orient.

There is, incidentally, not total agreement that the spiny greasebush belongs to the bittersweet family. It may really belong to another family, the Crossomataceae, otherwise known as the crabapple bush family. This family, which presently contains only three genera, is restricted to the deserts of southern Nevada, California, and Arizona. It appears to be distantly related to the magnolias.

There are several reasons why botanists have been uncertain on this issue. For one, detailed anatomical, biochemical, and physiological studies which might help decide the relationships of the spiny greasebush have not yet been made. But another and much more fundamental problem when dealing with plants such as this, which have relatively simple flowers, is that so many parts have been lost in the course of evolution that we frequently have to make what amounts to a shrewd guess regarding their possible ancestors. Several lines of evidence indicate that the earliest flowering plants of about 130 million years ago had many floral parts—many sepals, many stamens, many petals, etc.—and that their flowers were relatively undifferentiated. This pattern would be very much like that of the present-day magnolias. Evolution in the flowering plants, then, to a great extent, has involved the reduction or even the loss of various parts and the increasing specialization of those floral structures that remain.

RHAMNACEAE
BUCKTHORN FAMILY

Tobacco Brush
Ceanothus velutinus

INDIANS ONCE USED the leaves of *C. velutinus* as a substitute for tobacco, presumably a poor one, since it is not so used today. Other common names for tobacco brush include snowbrush, in reference to its soft mounds of white flowers in the spring; sticky laurel or varnish leaf ceanothus, because of the sticky, varnishlike coating on the leaves; and mountain balm, since this shrub inhabits montane situations and has a strong characteristic odor, leaving the subjective impression of cinnamon, balsam, or walnut. There is some evidence that its distribution is dependent on an insulating snow cover during the winter, so the term snowbrush would also be appropriate from this standpoint. In any event, this is the only *Ceanothus* with a significant distribution within the Great Basin, though many more species occur in California and the Southwest.

In a region replete with tough plants, tobacco brush has a cast-iron character. It rarely shows any evidence of grazing or insect depredation, and it is one of the first shrubs to appear after a severe burn. Even though the top of the plant may be burnt to the ground, tobacco brush readily root-sprouts. In addition, the seeds will lie dormant in the soil for a long time until the heat of a forest fire stimulates them to germinate. It has been shown that *Ceanothus* seeds immersed for a time in hot water will have improved germination rates. Experiments by J. R. Sweeney of the University of California have demonstrated that seeds of California buckbrush, *C. cuneatus*, will withstand dry heat up to 80 degrees C. and germinate as well as those of a control group kept at a lower temperature.

Clarence and Alice Quick, of the California Forest and Range Experiment Station, exposed the seeds of two species of *Ceanothus* to boiling water for varying periods of time. They found that, after being boiled for twenty minutes, 12 percent of the seeds of deerbrush, *C. integerrimus*, germinated. Seeds of mountain whitethorn, *C. cordulatus*, were even more resistant. The Quicks obtained a germination rate of 25 percent after mountain whitethorn seeds had been boiled for twenty-five minutes. Thirty minutes, however, was apparently too long, since none of the seeds germinated after being treated that long. Tobacco brush and other species of *Ceanothus* have a very hard seed coat which must be cracked or abraded in some way or else exposed to heat before germination will occur. Even then, germination will not occur unless the seeds are stratified, that is, moistened and stored for a few weeks at temperatures near 0 degrees C. Heating or breaking the seed coat will allow the seeds to imbibe water and swell, but growth will not take place unless they have been exposed to this sort of cold period.

The optimum time for exposure to boiling water appeared to vary with the particular sample of seeds. In one case, ninety seconds seemed not to be long enough, although over 95 percent germination was achieved. But in other cases five seconds of exposure produced the best results, with around 80 and 60 percent germination. This would indicate the existence of significant genetic differences, resulting in either ecotypes or biotypes. One might assume from this that some deerbrush populations are adapted to hotter fires than others. The Quicks also found that *Ceanothus* seeds remain viable for a very long time. One batch of deerbrush seeds twenty-four years old showed a 90 percent germination rate, while 92 percent of a tobacco brush seed sample which was twelve years old germinated.

The successful pioneering efforts of tobacco brush can be seen on the east slope of the Sierra Nevada, along Interstate 80 where it traverses the site of the infamous Donner Lake fire of 1960. The slopes, once covered with fir, Jeffrey pine, and lodgepole pine, are in many spots now solid stands of tobacco brush over a meter high and so impenetrable that even the hardiest hiker avoids them. The Forest Service has recently undertaken a program of spraying some of these areas with herbicide in order to open up sites for tree planting, so persistent is the tobacco brush. Obviously it is of considerable value, despite its unpalatability, because of its ability to rapidly invade a burnt-over area and as a result hold soil in place against what might otherwise be the disastrous effects of erosion.

Like the bitterbrush, the roots of tobacco brush have nodules on them

Tobacco Brush

which contain a type of fungus parasite known as an actinomycete. This is, in reality, a symbiotic relationship, since the fungus is able to fix free nitrogen from the air in return for the food it gets from the host plant. This nitrogen is chemically combined in a form which can be utilized by the tobacco brush and other plants. Although the rate of nitrogen fixing is probably not equal to that of root nodules on members of the pea family, it is probably safe to say that the amount of nitrogen fixed in this way by the root nodules on bitterbrush and tobacco brush is very significant on western ranges, though no precise figures are available.

Tobacco brush is evergreen, with broad, oval leaves from 4 to 7 centimeters long. Growing alone, it forms a round-topped shrub and may eventually attain a height of 2 meters. Like those of most other species of *Ceanothus*, the leaves have three particularly prominent veins arching upward from the leaf base. Initially the leaves are light green, but eventually they become dark green. They are smooth and shiny above and finely pubescent on the pale green undersurfaces. Frequently they are strongly curled toward the undersurface, especially so during dry or cold weather. The slightly pubescent young twigs are deep green, varying to brown for older branches. The white flowers are formed in dense clusters near the ends of the branches; each flower has five petals and five sepals united into a cuplike structure which is fused with the base of the pistil. The fruit, a seed pod or capsule 6 millimeters across, is distinguished by three prominent lobes at the top. Each lobe contains a single seed.

J. E. Keeley and P. H. Zedler carried out a study on fruit and seed production in two species of manzanita and two species of *Ceanothus* in San Diego County, involving both burnt and unburnt sites. One species of *Ceanothus* was a nonsprouter—that is, it did not come back from root sprouts after a fire—while the other species was able to reproduce from both seeds and root-sprouting. The nonsprouting *Ceanothus* produced more seeds and germinated more seedlings after a fire than the sprouting species. Intuitively, this is about what one might expect, since the nonsprouting species must depend entirely on seeds to come back after a fire, which usually destroys all the adult plants. We have to be careful not to be teleological in our interpretation, however: the nonsprouter, of course, does not anticipate the fire but, over the eons, under the pressure of natural selection, it has evolved this mechanism of greater seed production, a factor which insures its continued survival.

A clear example that intuition doesn't always serve as a reliable guide in instances of this sort was evidenced in the two manzanita species. Keeley and Zedler found that the nonsprouting manzanita produced fewer seeds than the sprouting species, but the seeds from the nonsprouter were larger and better protected, with the result that they stood a better chance of surviving a fire. Consequently, even though fewer seeds were produced, the nonsprouting manzanita showed more seedlings appearing after a fire because of its superior packaging. Seed production varied considerably from year to year in both the manzanita and the *Ceanothus* species, appearing to depend on the amount of precipitation during the preceding year. More precipitation meant more photosynthesis and thus more food which could be diverted into seed production the next year. In summary, we can say that these two *Ceanothus* species and the two manzanitas illustrate four different evolutionary strategies for coping with frequent fires, each of them apparently successful.

Tobacco brush is virtually valueless as forage for cattle and horses. Deer and elk will eat it, but only if better forage is not available. One experiment conducted near Truckee, California, involved ninety head of goats allowed to browse in a tobacco brush area for a period of five years. The goats remained in good condition, and some 60 percent of the tobacco brush was killed. However, another experiment near Mount Lassen produced contrary results, with the goats nearly starving to death in a manzanita and tobacco brush community.

Tobacco brush is common on relatively dry mountain slopes from 3,500 to over 10,000 feet. Its range continues northward from the southern Sierra Nevada in California to British Columbia. To the east it ranges through the Rocky Mountains as far as South Dakota.

The mountain whitethorn, *C. cordulatus*, mentioned earlier is found at the extreme western edge of the Great Basin and on into California. It is distinctive and easily recognized, completely unlike any other high-altitude species of *Ceanothus*. The grayish aspect, smooth, whitish bark, and rigid, intricately branched stalks frequently ending in sharp spines, unlike any of the other Great Basin species in the genus, make this shrub simple to identify, even from a speeding car. The pale leaves are 1 to 2 centimeters long, more or less elliptical in shape, and have three prominent parallel veins. The clusters of white flowers, produced during the summer, are followed by the characteristic three-lobed seed capsules of the genus.

The mountain whitethorn may attain a height of 2 meters, though it is usually much less. Generally the bush has a flattened appearance, apparently caused by the weight of heavy winter snows in the Sierra Nevada. Although not very high, some examples will have a branch spread of 4 meters. Plants in poorer sites tend to be much thornier. As a browse plant, this species is considered only poor to fair for cattle and sheep.

Mountain whitethorn extends from the Cascade Mountains south to Baja California and east only to extreme western Nevada. The species name is a Latin term for heart-shaped, although the leaf blades hardly fit that description. They are, however, sometimes indented at the base where the leaf stalk joins, and that is enough to call them cordulate.

Ceanothus is a member of the buckthorn or jujube family, Rhamnaceae, which is worldwide in its distribution with around fifty-eight genera and nine hundred species. About forty species of *Ceanothus* occur in California, many of them essential components of the chaparral vegetation. The genus name comes from the ancient Greek name for a variety of prickly plant, *keanothus*. The species name for the tobacco brush means silky pubescence, in reference to the surface of the underside of the leaves. There is a variety of *velutinus*, called *lorenzenii* in the past, with smaller leaves which are very little varnished above. This variety is now believed to be a hybrid with mountain whitethorn, C. *cordulatus*; the correct way to designate this is C. X *lorenzenii*. Another variety, without the velvety pubescence on the underside of the leaves, occurs in the Pacific Northwest and is known as *laevigatus*.

Sierra Coffeeberry
Rhamnus rubra

THE COFFEEBERRY or buckthorn is a member of a large genus of about one hundred species distributed primarily throughout the temperate and warm climates of the northern hemisphere. Perhaps the best known is the common buckthorn, *R. cathartica*, of the Old World. Found throughout much of Europe, North Africa, and northern Asia, this species produces berries which for over a thousand years have been the source of a strong cathartic and laxative, so potent that the flesh of some animals who have consumed the berries is said to retain the purgative properties. The juice pressed from the berries is used to prepare a "syrup of buckthorn." The bark and dried berries have been used as a source of yellow and saffron-colored dyes; the berry juice, when combined with alum, produces a green dye, once used by watercolor artists. Cascara sagrada, *R. purshiana*, native to California and the Northwest, which produces the same medicinal effects but to a milder degree, in the last century displaced the Old World common buckthorn in the pharmacopoeias of both the United States and Europe, along with the milder alder buckthorn, *R. frangula*, of Europe.

The sierra coffeeberry just barely gets into the Great Basin along its western edge. It can be found in the canyon of the Truckee River near Verdi in sparse stands, which the uninitiated may mistake for bitter cherry. There are several ways to tell these shrubs apart, however. While the twigs may be reddish in both and the small, elliptical to oblong, blunt leaves are about the same size—from 1 to 3 or 4 centimeters in length for the coffeeberry—bitter cherry leaves are characterized by one or two small, yellowish, round glands, discernible with the eye or a low-power hand lens, near the junction of the leaf blade and leaf stalk. These glands are absent in the coffeeberry.

Sierra Coffeeberry

Additionally, the leaves of coffeeberry are noticeably thicker, with very prominent parallel, lateral veins running out to the edge.

When in bloom, these two shrubs are easily distinguished. Bitter cherry has prominent white flowers, a centimeter in diameter with many stamens. The inconspicuous greenish flowers of the coffeeberry have only four or five stamens, and in this species five very small, bilobed petals appear to partially enclose the stamens; some species of coffeeberry have unisexual flowers and have completely lost their petals. Eventually, coffeeberry produces black berries about 8 millimeters long with two nutlets. Bitter cherry has a similar, slightly larger berry which turns from red to black but which has only a single small pit, like that of other cherries.

The sierra coffeeberry gets to be about 1 to 1.5 meters tall; it seems to prefer dry, open slopes in coniferous and sagebrush areas. The genus name *Rhamnus* comes from an old Greek name for buckthorn. The thorns, which, incidentally, do not occur in our species, are simply the tapered ends of some of the branches. The species name *rubra* refers to the slender, reddish branchlets.

ACERACEAE
MAPLE FAMILY

Dwarf Maple
Acer glabrum

EAST OF THE Great Basin, the dwarf maple is known as the Rocky Mountain maple, and in northwestern Washington it is hardly dwarf, reaching a height of 12 meters. As we might expect—with their propensity for debating both major and minor differences—botanists have considered this taller form to be a separate species, but most expert opinion today regards it simply as a variety, which has been named *douglasii*. Within our area, however, the dwarf maple is typically a 1.5- to 5-meter-tall shrub confined largely to the mountains; sometimes it is especially abundant along streams and in the wetter canyons. Commonly, it is found growing on rocky or gravelly soils in association with such other species as aspen, birch, red fir, and ponderosa pine. On drier sites its companions may be chokecherry, serviceberry, and snowberry.

The leaves of the dwarf maple vary between 2.5 and 7.5 centimeters long and are about as broad; their three prominent lobes are toothed along the edges. The species name *glabrum* refers to these smooth or hairless leaves, which are bright green above and somewhat paler beneath. In common with all maples, two leaves are attached at each node, an arrangement termed opposite by botanists. The main stems and older branches have a smooth gray bark, while the younger branchlets vary from green to reddish. Sometimes the leaves are so deeply lobed that they are divided into three leaflets, particularly on plants growing in shady and moist locations. This form was once described as a separate species called *tripartitum*; however, it is not now considered to be even a valid variety, since, not uncommonly, such forms will have both three-part or compound as well as simple leaves on the same branch.

One explanation of this difference in leaf shape may come from the recent studies of David Steingraeber on the formation of leaves in the sugar maple. In common with those of many other woody plants, the terminal buds formed during the current summer contain the preformed but unexpanded leaves that will appear next spring. These leaves are called early leaves. However, some shoots will subsequently produce additional or late leaves, which immediately expand without undergoing the usual winter dormancy period. Steingraeber found that these late leaves had deeper sinuses between the lobes and broader angles than the preformed leaves; he interpreted this as an adaptation that enabled these leaves to cast a smaller shadow on the leaves below, with the result that the multilayered canopy of the sugar maple was able to photosynthesize more efficiently. In a later study reported on as an abstract during the summer 1984 meeting of the Botanical Society of America, Steingraeber showed that essentially the same phenomenon occurs in the dwarf maple, with the late leaves sometimes being divided into three leaflets.

At the time that the leaves are produced in the early spring, clusters of greenish yellow flowers appear. A close examination of these will show that there are three types: one with both stamens and pistils, one with only stamens, and a third type with pistils and abortive stamens. These latter two types are sometimes the only kind present, and in that case they are borne on separate plants—that is, each individual has only staminate flowers or pistillate flowers. Since these are present on separate individuals, every seed produced will inevitably be the result of cross-pollination. The advantages of cross- versus self-pollination have been considered in our discussion of the smokebush. What is apparent here is evolution at work—come back in a few hundred thousand years or more and the dwarf maple will perhaps be capable of producing only unisexual flowers and no "perfect" ones. Whether this will be the case depends on how much of an advantage is conferred by cross-pollination. Cross-pollination always means more variation and very probably a more adaptable species.

The single pistil of either the pistillate or the perfect flowers produces two seeds, each enclosed in one-half of the pistil. The opposite outer walls of the pistil develop a thin wing several times longer than the basal portion which encloses the seed. When mature, the dry fruit splits into two winged affairs, each enclosing a single seed. When they fall from the tree, these one-bladed propellers rotate in a fashion unique among those natives which produce

Dwarf Maple

winged fruits. Of course, if there is wind, the seeds may be carried some distance from the parent shrub. Winged seeds of this sort, called samaras, obviously confer a considerable advantage by allowing the rapid spread of their species.

Two varieties of the dwarf maple are sometimes recognized in the western Great Basin. *A. glabrum* variety *torreyi*, found in western Nevada and the Sierra Nevada, is characterized by its usually reddish twigs, while *A. glabrum* variety *diffusum*, found on the east slope of the Sierra Nevada, has grayish twigs. This latter variety is the most common form throughout the Basin.

Much of the brilliant fall coloration in eastern deciduous forests is associated with the various species of maple, which produce colors ranging from yellow to orange and scarlet. Our dwarf maple appears to be precocious in its development of a bright scarlet color on some leaves during the middle of the summer. In reality, this is due to a fungus, which infects the leaves, and sometimes to an insect injury, which interrupts a leaf vein. When the latter damage takes place, the ability of the vein to transport the sugar produced by the leaf is interfered with, and the excess sugar is converted to a red pigment. Apparently, the fungus can produce the same end result. This process is basically very similar to those events which take place in leaves that become brightly colored in autumn, except that then the interruption of sugar transport occurs as a result of a corky layer which develops at the point of attachment of the leaf to the stem and effectively cuts off the transfer of sugar out of the leaf. This corky layer, in turn, is stimulated to form by typical football weather—warm, sunny days and cold nights. So, fall colors are really the result of plants getting ready to shed their leaves in preparation for winter. Oddly enough, in the fall the dwarf maple turns golden yellow rather than red, as might be expected. As far as anyone knows, the only beneficiary of the bright colors of autumn are humans with the aesthetic sensibilities to appreciate them!

The genus name *Acer* is the old Latin name for the maple tree. There are over 110 species in the genus, most of them concentrated in China. The fossil record indicates that the maples evolved at least 70 million years ago, near the end of the age of dinosaurs. Most maples are confined to the north temperate zone; the family to which they belong, the Aceraceae, contains only one other genus, also centered in China and containing only two species. Close relatives are considered to be the horse chestnut and tropical litchi nut families. In the eastern deciduous forests of North America, the sugar maple is considered to be one of the final dominant tree species, along with the beech. That is, after a succession of various species, the forest tends to be dominated by an association of beech and sugar maple. For this reason, beech and sugar maple are sometimes called climax species.

ANACARDIACEAE
SUMAC FAMILY

Squawbush
Rhus trilobata

A LIST OF the other common names of the squawbush is a good indicator of some of its more obvious characteristics: three-lobed sumac, polecat bush, skunkbush, squawberry, and lemonade sumac. As might be surmised from all this, the squawbush has a disagreeable odor, three-part leaves, and acid fruits. The name squawbush refers to its use by the Indians in basket making and in medicine. Usually its leaves are composed of three leaflets rather than three lobes, even though the Latin species name implies that three lobes are characteristic. The name sumac is a general one for members of this same genus. The family to which it belongs is also referred to as the sumac family, though this is an appropriate name only in the temperate zone.

Squawbush typically occurs on dry, rocky slopes, on cliff faces, and sometimes in moist valley bottoms. One unusual form is found on sand dunes just to the east of Delta, Utah. Although common from Baja California to the western foothills of the Sierra Nevada and extending northward into Oregon, it is absent from the extreme western Great Basin. Outside of the basin, in southern Nevada, it is abundant, and from there it extends eastward and northward into White Pine and Elko counties and on into Utah. Its total range includes the Rocky Mountains and areas east to Iowa and Texas.

Generally between 1 and 2 meters tall and spreading to an equivalent diameter, squawbush is an attractive shrub, with smooth brown bark and shiny, bluntly toothed leaves that are usually made up of three leaflets, 1 to 3 centimeters long, which are wedge-shaped at the base—although sometimes the leaves are three- or even five-lobed, rather than being divided into leaflets. The small, yellow flowers are produced in clusters at the ends of the branches early in the spring, before the leaves appear. The flowers are built on a very regular pattern of fives, with five sepals, five petals, five stamens,

Squawbush

and one pistil. After pollination, the pistils mature into red, sticky fruits about 6 millimeters in diameter. The structure of the fruit resembles that of the chokecherry, inasmuch as the flesh portion surrounds a pit which in turn encloses the seed.

Various Indian tribes throughout the range of the squawbush made extensive use of its berries. One popular use involved boiling them in water to produce a pink "lemonade." The berries can be eaten either raw or cooked, or they may be dried and kept for later use. They were sometimes ground up and made into cakes. The supple branches and bark were used extensively for baskets and other wicker ware.

Generally, squawbush has a low browse rating for cattle and sheep, although it appears to be more palatable toward the southwest, particularly so for goats. Even for deer it is considered only fair to poor. Oddly enough, its relative, poison oak, is considered to be more palatable to wildlife and cattle—horses will feed heavily on it, and deer prefer it to most other shrubs. For these animals, poison oak causes none of the allergic skin rash problems that it is noted for in humans and domestic pets.

Poison oak, *Toxicodendron diversilobum*, or *Rhus diversiloba*, as it was formerly known, is extremely common in some areas of California, especially in the foothills of the Coast ranges and the Sierra Nevada. It extends both into Baja California and north into Washington. Logically, it might be expected to occur in the moister sections of the Great Basin. A student in one of the author's classes, Bill O'Donnell, found a small stand of poison oak in a canyon near Pyramid Lake in the fall of 1983. As far as we know, this is the only example yet found in the Basin. Although there are reports of its eastern relative, poison ivy, *Toxicodendron radicans*, from Elko County, Nevada, these have yet to be confirmed. Extensive search in that area by the author has failed to turn up any examples. Poison ivy does occur to the east in Utah, in the mountains bordering the Great Basin.

Poison oak and poison ivy, like the squawbush, have leaves composed of three leaflets, but these are much larger—2.5 to 10 centimeters long—and thinner. The leaf stalk or petiole will get to be between 1 and 10 centimeters long, while that of the squawbush is only 1 to 2 centimeters in length. The easiest way to separate these species, however, is by means of berry color—poison oak and poison ivy always have white berries, while those of the squawbush are always red. Poison ivy, despite the name, grows commonly as a shrub or sometimes climbs; it is not really ivy- or vinelike. In fact, all sumacs that are poisonous to touch, such as poison ivy and poison oak, always have white berries, and the nonpoisonous ones are consistently red. The box elder, in the maple family, is sometimes mistaken for poison oak because many of its leaves are composed of three similar blunt-toothed leaflets. It can be differentiated very easily, however, by the fact that there are invariably two leaves attached opposite one another at each node. Poison oak has only one leaf at each node, a so-called alternate arrangement.

Both poison oak and squawbush are extremely variable, so much so that some authorities have divided them into a number of species. These have not stood the test of time, and most have been considered invalid or reduced to the level of varieties.

Another sumac which skirts the Great Basin is the smooth sumac, *R. glabra*, found in Utah in the foothills of the mountains. This species extends into the Pacific Northwest and down into California at one isolated locale; it is common in the East, ranging from Nova Scotia to Florida. Smooth sumac is nonpoisonous and has leaves composed of four to eight pairs of leaflets, with a single leaflet at the tip. Its red berries are produced in a large, elongated, terminal cluster.

The sumac family, the Anacardiaceae, is mainly tropical, with about 80 genera and 600 species. The cashew, pistachio, and mango are cultivated tropical members. The family is probably closely related to the maples and horse chestnuts. The genus *Rhus* contains about 120 species distributed over the world—one found in Japan, *R. verniciflua*, is the source of lacquer. The genus name for squawbush comes from the ancient Greek *rhous*, sumac. The word sumac apparently is an altered Arabic word; sometimes it is spelled sumach.

SOLANACEAE
TOMATO FAMILY

Shockley's Desert Thorn
Lycium shockleyi

THERE ARE A fair number of desert thorn species in the American Southwest, perhaps about ten or twelve, all of them shrubs. The uncertainty occurs because some disagreement exists about the validity of a few species. At any rate, the family to which they belong, the Solanaceae, commonly known as the nightshade family, has most of its members in the tropics; it has not been very successful in colonizing desert areas over the world, except for the genus *Lycium*, which contains about one hundred species worldwide.

During the cooler seasons, when the shrubs are leafless, it is difficult for the uninitiated to distinguish Shockley's desert thorn from black greasewood or Bailey's greasewood. But, if enough time is spent hiking over our Great Basin deserts and observing the distinctive characteristics of each shrub species, the differences, though subtle, will allow the desert thorn to be recognized even from a distance. For one, it is larger than Bailey's greasewood, and the ultimate branchlets are stouter and somewhat more angular. Black greasewood, although the same size or smaller, is almost always located in the dense, saline soils of alkali flats or playas, while desert thorn tends to grow in gravelly, better-drained areas, such as alluvial fans and foothill slopes. Frequently it will be found associated with Bailey's greasewood, but usually in a transitional area, where the slope becomes steeper and some of the other vegetation attests to a somewhat less dry environment.

However, even in the middle of winter, both Bailey's and black greasewood will usually still have a few dry fruits remaining from the previous summer. These resemble saucers with a pointed projection in the middle and are so distinctive that they can be used as immediate recognition characters for greasewood. Of course, if no fruits are present, one has to go by general appearance in the leafless period.

Shockley's Desert Thorn

During the spring and early summer, the fleshy, 1- to 3-centimeter leaves of Shockley's desert thorn are evident, and in their wider, spatulate appearance they are distinctively different from those of the greasewood. By midsummer the leaves have turned yellow and have begun to fall, in common with those of many other desert shrubs. This early leaf fall is characteristic of those desert shrubs which possess only a so-called C_3 type of photosynthesis. Without going into details here, we can note that plants with C_3 photosynthetic ability are efficient food makers only at lower light intensities and temperatures, so it is not surprising that they have adapted to the severe conditions of midsummer by simply shedding their leaves.

The small, light green flowers of Shockley's desert thorn are around a centimeter long. In common with those of many members of the nightshade family, the flowers have five sepals fused together to form a lobed cup, while the five petals are also fused into a funnel-shaped corolla with five lobes flared outward at the top. All the desert thorns produce berrylike fruits, but in some the pulpy portion dries at maturity to produce hard, woody fruits. In the Shockley's desert thorn, the originally yellow-green fruit becomes constricted above the middle and soon dries to form a hard fruit.

Fernando Chuang at the University of Texas, who recently studied all our southwestern members of the genus, concluded that much of the material which we once called Cooper's desert thorn, *L. cooperi*, is in reality Shockley's. A major difference between the two species involves the corollas—in Cooper's they are hairless, but in Shockley's they are obviously hairy. John Kartesz, who is presently working on a flora of Nevada, agrees with this assessment. For a long time, the name *shockleyi* was regarded as a synonym of *cooperi*. Both these species, incidentally, were described by Asa Gray in the 1880s. Gray, a professor at Harvard for most of his career, eventually gained a reputation as America's greatest botanist. He published many manuals and reports on the plants collected on various expeditions to middle and western America. Gray attempted to write a flora of North America, but this project proved to be impossible because of the plethora of new species described every year.

Cooper's desert thorn, known from southern California and Mexico, extends north into western Nevada as far as Esmeralda County and into extreme southwestern Utah. Shockley's desert thorn is a more northern species, found in the western Great Basin but extending beyond it to southern

Nevada. In eastern Nevada and western Utah another species, *L. andersonii*, is easily distinguished by its cylindrical, fleshy leaves up to 15 millimeters long, whitish or somewhat lavender flowers with a narrow corolla, and fleshy, bright red berries. Anderson's desert thorn can apparently grow in somewhat more saline soils than Shockley's. The matrimony vine, *L. halimifolium*, occasionally escapes from gardens in the Great Basin; it has purple flowers and slender, climbing stems. A very closely related species, *L. barbarum*, also appears to be an occasional escape.

Those species of *Lycium* with fleshy berries were used by the Indians, who either ate the berries fresh or dried them after boiling for later use. Other species to the south of ours appear to be of some use as a livestock browse. The berries are also important to wildlife. Other common names for various species in the genus include wolfberry, chico, squawthorn, and rabbit thorn.

Asa Gray named *shockleyi* in honor of the collector who first found this particular species near Candelaria, Nevada. Anderson's desert thorn was also named by Gray—for Charles Lewis Anderson, a physician-botanist who collected this shrub in southeastern Nevada. Anderson came to Nevada from Minnesota and set up practice in Carson City in 1862. He collected actively over the state for five years, and many of our Great Basin plants bear his name, most notably the desert peach, *Prunus andersonii*.

The nightshade family, to which this genus belongs, is a large, cosmopolitan one with around ninety genera and three thousand species. Most of the species are concentrated in Australia and South and Central America. Some economically significant members of the family are tobacco, potato, tomato, pepper, and belladonna.

POLEMONIACEAE
PHLOX FAMILY

Prickly Phlox
Leptodactylon pungens

THE GREAT BASIN prickly phlox, sometimes known as granite-gilia, barely qualifies as a shrub, since it is woody only at the base, although other species in the genus outside of the Basin are more obviously woody. It occurs in a variety of habitats, but generally it is found in dry and rocky places, from 4,000 to 12,000 feet in elevation. Typically, it is low-growing, getting no higher than 30 centimeters and sometimes forming mats at least as wide. The prickly aspect comes from its individual leaves, the tips of which are rigid and sharp-pointed. The leaves, divided into three to nine narrow segments, are closely spaced on the stems. The flowers are truly phloxlike, with a narrow, tubular base flaring out into five lobes at right angles to the tubular portion. Generally the flowers are white in color, although pinkish or yellowish forms are sometimes found.

A flower form with a narrow tube or throat, such as that of the phlox, favors an insect pollinator with a long proboscis. This kind of narrow throat also represents a protective device against robber insects, such as ants and beetles, which feed on pollen but are, for the most part, not good pollinators because of their smooth bodies and because, unlike bees, they do not become conditioned to visit only a particular type of flower at a given time. One study, at least, indicates that butterflies are the most frequent pollinators of phlox. Another study shows that slight color differences in phlox flowers are recognized by pollinators and that some pollinating insects show a flower constancy by visiting only a particular type of flower at a particular time. This behavior pattern helps maintain the genetic integrity of various species, since it reduces the probability of hybrids. This species of prickly phlox was once considered to have about five subspecies or varieties.

Prickly Phlox

So many intergrades are known, however, that the wisdom of recognizing several of these varieties is in doubt.

Like that of most other members of the phlox family, the fruit of the prickly phlox is a three-valve capsule which splits open at maturity to release the small seeds. Even though the phlox family is well represented within the Great Basin, only the prickly phlox is significantly shrubby; however, several species of the genus *Phlox* are somewhat woody at the base.

The genus name *Leptodactylon* comes from two Greek words, *leptus* meaning narrow and *dactylon* meaning finger. The species name *pungens* comes from the Latin and means puncture. The phlox family, the Polemoniaceae, is composed of about eighteen genera and three hundred species. Western North America has most of the species in the family, although it extends down to western South America and north through Alaska to Asia and Europe.

LAMIACEAE
MINT FAMILY

Purple Sage
Salvia dorrii

I ONCE ASSUMED that *Riders of the Purple Sage*, by Zane Grey, referred to our desert or purple sage. To my chagrin, I later learned that this famous western author was simply guilty of a little artistic license in calling our common sagebrush, an entirely unrelated plant, purple sage. Purple sage is a true sage, since it belongs to the same genus as the cultivated *Salvia* used in the culinary arts. Like the latter, the purple sage has the strong, aromatic, characteristic sage fragrance. In one species of California sage, it has been found that the volatile terpenes produced by the leaves appear to interfere with the germination and growth of grasses. This kind of competitive chemical behavior by plants is known as allelopathy.

The purple sage in the Great Basin characteristically grows on dry slopes and rocky bluffs in sagebrush associations. Except when in bloom, it is seldom abundant enough to attract much attention. Ordinarily it appears as a low, rounded shrub, seldom taller than 60 centimeters. The gray, finely pubescent, and rounded leaves—1 to 2 centimeters long—persist to some extent through the winter. The blue flowers are borne in erect, interrupted spikes, with conspicuous bracts colored pink to purple surrounding them. Like the other members of the mint family, to which it belongs, purple sage has two-lipped flowers.

Botanists recognize two basic types of flowers with regard to the symmetry of petals and sepals. If a flower has an arrangement like that of the lily, in which all the sepals and petals are alike, and the flower is essentially circular in outline, it is referred to as being regular or actinomorphic. Such symmetry is regarded as radial. On the other hand, if a flower has the kind of symmetry exemplified by an orchid, it is regarded as irregular or zygomorphic. In such

Purple Sage

flowers there is only one plane of symmetry—there is only one way to split the flower, with one half looking like the mirror image of the other half.

Irregular flowers are somewhat better adapted for specialized insect visitors, and many forms can be effectively pollinated by only one kind of insect. Many have evolved ways to insure that any visiting pollinator will successfully transfer pollen from the stamens of one flower to the receptive stigma on the pistil of another flower. This complex arrangement ultimately makes for a more efficient system in terms of the number of seeds produced compared to the number of pollen grains formed. One could say that it was very energy-efficient, at least for the plant involved, although the insect obviously expends a lot of effort by flying to various flowers in order to collect pollen and nectar. This increased efficiency does not necessarily mean that such irregular flower types will eventually take over the plant world. One has only to consider the very successful grasses, all of which are wind-pollinated.

In sage flowers, the upper lip is composed of two fused petals, while the lower comprises three fused petals. There are four stamens, but only two of these are fertile. The supporting stalk for each stamen is fused at the base with the corolla, a collective term for the petals. In the young flower, the two fertile stamens are arched over in such a way that a bee attempting to get at the nectar at the base of the flower will force the anthers of the stamens to dust its back with pollen. At this stage, the receptive stigma is held high, out of the bee's way, under the arch formed by the upper lip of the flower. After the pollen has been shed, the stalk supporting the stigma elongates and arches over into a position which allows it to contact the back of any bee entering the flower. In this way, self-pollination is avoided and cross-pollination is assured. The offspring resulting from the latter are more variable and thus confer an evolutionary advantage on their species.

Like irregularity, fusion of the stamens and petals is considered an advanced evolutionary feature. Although not significant as such in the Great Basin, the various species of *Salvia* are important bee plants in California and thus important in the production of honey. In common with that of other members of the family, the species' pistil is a four-lobed affair. At maturity, these lobes split apart to form four nutlets, each containing a single seed.

Altogether, there are about 500 species of sage distributed over the temperate and warm regions of the world. The mint family or Lamiaceae contains about 200 genera and around 3,200 species, distributed in almost all habitats from the Arctic to the tropics. The greatest concentration of species occurs around the Mediterranean Sea. Only the tropical rainforests have relatively few members of this family.

Salvia comes from the Latin verb *salveo*, meaning to save, in reference to this species' medicinal properties. Several varieties of the purple sage are recognized. The typical form occurring through much of the Great Basin is variety *carnosa*, with leaves 15 to 30 millimeters long. In the southern Basin another variety, *argentea*, occurs: this form has smaller leaves, 4 to 15 millimeters long. It differs from the typical form of the species, variety *dorrii*, in that the latter has floral bracts which are long and hairy, while those of the former are generally nearly hairless, except for a row of hairs along the margins. Variety *dorrii* is found in southern Nevada, California, and Arizona, essentially outside the Great Basin.

CAPRIFOLIACEAE
HONEYSUCKLE FAMILY

Twinberry
Lonicera involucrata

THE TWINBERRY is another of our shrubs that just skirts the Great Basin. In California, it extends from sea level to 9,500 feet and occurs on the eastern slope of the Sierra Nevada. Twinberry extends north into Alaska, east across Canada to Quebec, and south through Colorado and Utah to New Mexico and Arizona and on into Mexico. Wherever it grows, it prefers moist, humid locales, and perhaps for this reason it does not get far into the Great Basin.

The deciduous leaves and stems of twinberry are paired at the nodes, and the individual leaves are oval and pointed, dark green above and somewhat paler and finely pubescent beneath. They vary in length from 5 to 12 centimeters. A somewhat lax shrub, twinberry is generally around a meter tall, but sometimes, in favorable locales, it grows to 3 meters. Individual flowers are borne in pairs on a long stalk. Each pair of flowers have at the base two bracts (really four, but not obviously so), which are partially united into a kind of cup or involucre. These bracts enlarge and turn red as the fruits mature. The flowers are yellow, sometimes with a reddish tinge. They are about a centimeter long, with five petals united into a cylindrical tube; nectar is secreted at the base of this tube in a shallow cup. A cylindrical, fused flower such as this helps the plant restrict its kinds of pollinators and insures more efficient pollination. Bees appear to be the most important pollinators of the twinberry, although hummingbirds are frequent visitors.

Another species of *Lonicera* which gets to the fringes of the Great Basin is *L. utahensis*, the Utah honeysuckle. This species differs from the twinberry in having blunt-tipped leaves only 2.5 to 6 centimeters long and white, or at most light yellow, flowers. Despite the name, it occurs in northern California, although not on the east side of the Sierra Nevada, and extends

Twinberry

Double Honeysuckle

through the Northwest and Canada into Idaho and then down into Utah. This particular species was first collected by Sereno Watson in 1871 in the "Wahsatch" Mountains, Utah, in Cottonwood Canyon at 9,000 feet.

Still another honeysuckle, found in western Nevada and particularly abundant in the Lake Tahoe area, is the double honeysuckle, *L. conjugialis*, in which the bracts below the purplish black flowers are very small, unlike the large bracts of the twinberry. The double honeysuckle also has much smaller leaves, 2 to 6 centimeters long. This is a plant of the Sierras and the Northwest, however; it does not adjoin the northern and eastern portions of the Great Basin, as do twinberry and Utah honeysuckle.

Twinberry obviously gets its name from the paired black berries, each about 8 millimeters in diameter, that mature from the fertilized pistils. Although one book on edible wild plants characterizes the berries as pleasant-tasting, most would agree that they are bitter or sour at best. Since the European species of honeysuckle produce berries which are definitely poisonous, and the fruits of all honeysuckle species are regarded as emetic and cathartic, they are better avoided in favor of something less dubious. As forage, twinberry appears to be of value only for deer and other wildlife.

Twinberry and the honeysuckles belong to the honeysuckle family, the Caprifoliaceae. The elderberries and snowberries also belong to this family, whose characteristics are covered under the discussion of the elderberries. The genus *Lonicera* gets its name from Adam Lonitzer, a German herbalist of the sixteenth century. There are about 150 species of honeysuckle, distributed primarily throughout the north temperate zone. The species name *involucrata* refers to the two united bracts beneath the paired flowers. The double honeysuckle, *L. conjugialis*, is so named from the two fruits which, unlike those of the twinberry, are fused together at maturity. The twinberry also goes by other colorful names: bearberry honeysuckle, fly honeysuckle, skunkberry, black twinberry, and inkberry.

Elderberry
Sambucus spp.

The seeds contained within the berries, dried, are good for such as have the dropsie, and such as are too fat, and would faine be leaner, if they are taken in a morning to the quantity of a dram with wine for a certain space. The green leaves, pounded with Deeres suet or Bulls tallow are good to be laid to hot swellings and tumors, and doth assuage the paine of gout. —
Gerard's HERBALL as quoted in M. Grieve, A MODERN HERBAL

THERE IS, perhaps, more mythology surrounding the elderberry than any other native shrub of the Great Basin. The stories, which originated with those species native to Europe, extend back to the ancient Greeks and Romans. One tale has it that Judas hanged himself from an elder tree. However, this is pretty improbable, considering the typical size of an elderberry; some scholars think that this legendary tree was actually a species of *Cercis* or redbud. Another medieval belief was that Christ's cross was made from elder wood. Some people considered that it was the abode of departed spirits and felt constrained to tip their hats when passing by an elderberry. M. Grieve's book is recommended to all who would explore some of the interesting, if fanciful, tales about this popular plant.

Great Basin Indians were aware of the therapeutic value of elderberry. According to the USDA contribution on medicinal uses of plants by Nevada Indian tribes, authored by Train, Henrichs, and Archer, a tea made from the flowers was used for tuberculosis as well as for colds and as a general tonic. The berries were used as a cure for diarrhea. The leaves were valued as a treatment for bruises and wounds. A root decoction was prescribed as a blood tonic and was supposed to stop dysentery.

There are two species of elderberry in the Great Basin. One is *S. cerulea*,

Elderberry

the blue elderberry, so-called because of its blue (sometimes black) berries. The waxy bloom of the berries, which is responsible for the blue color, occasionally extends to the young branchlets. The blue elderberry ranges throughout California, north to British Columbia, east to the Rocky Mountains and Arizona, and south into Mexico. The other species is *S. racemosa*, the red elderberry, which has bright red fruits. This form has an even wider range, growing north to British Columbia and east to Georgia and Newfoundland; the range of this species also extends to Europe. A variety of this species, *melanocarpa*, the black elderberry, has black fruits and also occurs in the Great Basin. This form can be easily distinguished from the black-fruited form of the blue elderberry by the shape of the fruit cluster. In the blue elderberry, the cluster is flat-topped, while the red elderberry, as well as its black form, has dome-shaped clusters.

The elderberries prefer moist locales, along streams or seepage areas in the mountains or sometimes on the north side of cliff faces, protected from excessive dehydration. Their lush green foliage is easy to recognize, even from a distance. Each leaf is composed of a main axis with five to nine leaflets; each leaflet is between 2.5 and 15 centimeters long and is finely toothed along the edge. Another obvious identifying characteristic is the occurrence of two leaves at each node or attachment point on the stem. Very few other native shrubs in our area have this kind of compound leaf borne two at each node. About the only other shrub with which the elderberry might be confused is the box elder; the latter, however, commonly has just three to five leaflets which are lobed and coarsely, rather than finely, toothed along the edge.

Elderberry flowers are white or cream-colored and small, only about 5 to 6 millimeters across. They have five very small sepals, five petals fused at the base to form a flat disk, five stamens alternating with the petal lobes, and a single pistil. The petals, sepals, and stamens appear to be perched on top of the ovary of the pistil, a condition known as inferior to botanists. Those families with inferior ovaries are considered to be advanced on the evolutionary scale. Primitive flowers have many separate parts, little distinction between sepals and petals, and superior ovaries. On this basis, the elderberry would also be considered advanced because of the fusion of its petals at their base.

Elder wood is quite hard and, but for the limited size of the trunk, would have been more generally used than just for durable small objects, such as combs, spindles, and pegs. Every country boy for centuries has known that the soft, easily removed pith of young elderberry stems was ideal for the construction of popguns and flutes. The berries are popular for wines, jellies, and sauces. The young vegetative shoots in the spring are edible if cooked, and even the blossoms can be eaten or made into wine. There are, however, reports that the fruits of the red-berried elders are sometimes toxic. The red berries frequently have a bitter taste which cooking does not dispel, and some people have suffered severe intestinal upsets after consuming them. Elder is considered as fair browse for cattle, but interestingly enough it appears to become very palatable to deer and livestock after the first heavy frosts have turned the leaves black.

Elderberries belong to the honeysuckle family, the Caprifoliaceae, which is rather small as flowering plant families go, consisting of about fifteen genera and a little over four hundred species. While it is worldwide in distribution, the greatest number of its species are concentrated in the temperate zones of eastern North America and eastern Asia. The family name comes from the Latin terms *capra*, or goat, and *folium*, which means leaf. The genus name *Sambucus* is derived from the Grecian term *sambuca*, a stringed musical instrument supposed to have been made from elder wood. The species name *cerulea* comes from the Latin for blue. The red elderberry, *racemosa*, gets its name from the shape of its fruit cluster. The Latin word *racemus* originally meant the stalk of a cluster of grapes; by metonymy, it eventually came to mean the cluster of grapes and even grape juice. In botanical terminology, racemose is a term used to refer to an *unbranched* inflorescence—not an altogether logical derivation!

Snowberry
Symphoricarpos spp.

THE SNOWBERRIES closely resemble their relatives, the twinberries. Like the twinberries, they may have their flowers borne in pairs. However, there is one easy way to distinguish them: twinberries always have red or black berries, while the snowberries have only white berries. When in flower, they can be separated by the fact that snowberries consistently have radially symmetric or regular flowers; that is, the petals are all the same size and shape. Twinberry flowers, on the other hand, always show some irregularity or bilateral symmetry, like that of the honeysuckles. Both genera have two leaves borne at each node, but the snowberry has smaller leaves, generally between 1 and 3 centimeters long, although sometimes on sterile shoots in the shade they may be somewhat larger.

Four species of *Symphoricarpos* occur commonly throughout the Great Basin. They are not always easy to separate from one another, and a few experts have shifted them back and forth between various names over the years. John Kartesz, as a result of his labors over the flora of Nevada, considers that our commonest species is *S. oreophilus*, the mountain snowberry, which is found in two forms, a smooth and a pubescent variety. This is an erect shrub which may get as tall as 1.5 meters, though it is generally much less. The tubular flowers, produced from spring until the end of summer, are about 7 to 9 millimeters long and cream-colored to pinkish in this species.

A similar species, which tends to be a lower, spreading form, is Parish's snowberry, *S. parishii*, which has somewhat smaller flowers 6 to 7 millimeters long. Both snowberries reproduce abundantly by vegetative means—the mountain snowberry develops many underground runners, while the procumbent branches of Parish's snowberry root easily at the tips. Both shrubs inhabit dry, rocky slopes and ridges from around 4,000 to 11,000 feet.

Snowberry

Another low form with arching branches that root at the tips is *S. acutus*, the creeping snowberry, which resembles Parish's except that its flowers are smaller, only 3 to 5 millimeters long, and bell-shaped rather than tubular. In addition, the creeping snowberry has densely pubescent leaves, while those of Parish's are only sparsely pubescent. Within the Great Basin, the creeping snowberry has been found as far east as Elko and White Pine counties in Nevada as well as south, outside our area, in Clark County, Nevada.

A species which occupies even drier sites on our desert mountain ranges is the desert snowberry, *S. longiflorus*. As the Latin name implies, it has longer flowers, some 10 to 15 millimeters long. These flowers are also characterized by an upper portion which flares out at an angle to produce a flower with a wider diameter than those of the previous three species. Tubular flowers which have an upper portion flaring out in this fashion at some angle to the long axis of the flower are described as being salverform; the word comes from the Latin *salvare*, which means tray. A phlox flower is a good example of this salverform type, since its upper portion resembles a tray with five scallops around the edge. Tubular flowers restrict the kinds of pollinators that can effectively visit a flower, and salverform types present a larger visible target for the insect as well as a landing platform, while the tubular basal portion insures that only the right pollinators will be rewarded. Unfortunately, all these species have somewhat similar oval leaves and are difficult to identify if only vegetative branches are available.

The fruits mature in the form of white, round or oval berries 5 to 8 millimeters long, each enclosing two nutlets. Because of the fruits' waxy appearance, another common name for snowberries is waxberry, and they have also been called Indian currants and wolfberries.

On many of our Great Basin ranges, the snowberry is a dominant shrub over large areas—it is fair to say that it is as characteristic of the mountain brush habitat as is big sagebrush in sagebrush-grass zones or shadscale in the salt deserts. Because of its abundance, it is an important browse plant for sheep. Even cattle and horses will eat it, although cattlemen will tell you that it is only a fair to poor plant for this purpose. Its value to deer is apparently about the same.

Like the elderberries and twinberries, *Symphoricarpos* is a member of the honeysuckle family, the Caprifoliaceae. Some of the characteristics of this family were discussed under our treatment of the elderberry. One thing we did not point out, however, was the fact that all members of the family always have two leaves at a node. One of the famous members of the honeysuckle family, *Linnaea borealis*, the twinflower, is found in the cool, northern coniferous woods of Asia, Europe, and North America. The genus was named after the famous Swedish botanist responsible for our present binomial system of naming plants, Carl von Linné or Linnaeus, as he is more commonly known. Many of our plants are named in honor of the famous or the obscure, although it is still considered bad form to name a plant after yourself! But, if you are the author of a new species, your name will be included after the species name as the "authority" for that species. However, the name and a description of the new plant must be published for it to be recognized as valid by other botanists. All this does not mean that your new species will be recognized as a good one, but at least you will have achieved immortality of a sort! As one of my botanist friends once remarked, "They may not agree with your new species, but they can never totally ignore you."

The genus name *Symphoricarpos* comes from the Greek and means clustered fruit. The species name *oreophilus*, also derived from the Greek, means mountain-loving. *Parishii* honors Samuel Parish, a pioneer botanist of southern California during the last half of the nineteenth century, and the Latin *acutus* refers to the pointed leaves of that species.

ASTERACEAE
ASTER FAMILY

Dwarf Sagebrush
Artemisia arbuscula

DWARF OR LOW SAGEBRUSH is much like big sagebrush in general appearance, but, as the name implies, it is much smaller, usually no taller than half a meter and typically much less. The easiest way to separate dwarf from big sagebrush is to look at the leaves—if they are more than three times longer than they are broad, the species is big sagebrush; if they are less than three times longer than broad, it is dwarf sagebrush. There are other differences, of course. Dwarf sagebrush tends to be a darker shade of gray, so much so that the variety (or subspecies, according to some authorities) *nova* is frequently called black sagebrush. Both black and dwarf sagebrush leaves have three lobes at the broad end, as do those of big sagebrush, but they have a much broader, wedge-shaped appearance, ranging in length from 5 to 15 millimeters. Sometimes the lobes are separated by deep divisions, so that they appear as fingerlike extensions.

Generally, dwarf sagebrush grows on rockier, poorer soil than big sagebrush. Very commonly, it will grow adjacent to big sagebrush communities, but there is typically a sharp transition from one community to the other. Quite often, what may appear to be a big sagebrush community of gradually decreasing stature will, on closer inspection, be seen to be a transition to a dwarf sagebrush community—the steeper hillsides, particularly if only a thin soil layer is present, will be covered with dwarf sagebrush. Frequently, these are degraded big sagebrush sites that have lost surface soil layers over geologic time.

Ascending our desert mountain ranges, we can observe that the coldest and driest woodland sites are frequently occupied by dwarf sagebrush, while black sagebrush usually occurs in locations which appear to have somewhat intermediate temperatures. Marda L. West, studying the big, dwarf, and black sagebrushes in the White Mountains of California, found that dwarf sagebrush occurred on dolomite, sandstone, and granite soils, while black sagebrush was restricted to limestone soils in pinyon-juniper woodlands. Dwarf sagebrush and big sagebrush did not appear to do well on dolomite soils. Otherwise, West found that big sagebrush was far more adaptable than either dwarf or black sagebrush. Big sagebrush in the White Mountains ranged from 6,000 to 10,800 feet through four major vegetation zones: shadscale desert, pinyon-juniper woodland, the subalpine community, and the alpine zone. Dwarf sagebrush was found between 10,000 and 12,800 feet within the two zones known as subalpine and alpine. Black sagebrush had the most restricted distribution of all—occurring only between 7,000 and 9,500 feet. Dwarf sagebrush is distributed from the Coast ranges of northern California through the Great Basin to southwestern Montana and northwestern Colorado. Black sagebrush occurs from the central Sierra Nevada of California east to southern Montana and south to southern California and northern New Mexico.

Big sagebrush has a considerable ability to become acclimated to a wide range of temperatures. Dwarf sagebrush also shows some ability to become acclimated, but West's studies showed no such capacity for black sagebrush. At least in the White Mountains, big sagebrush carried out photosynthesis most efficiently at 20 degrees C., while the comparable peak for dwarf sagebrush was 15 degrees C. Black sagebrush carried out food manufacture best at 25 degrees C. It is apparent that this differing efficiency at various temperatures is one of the major reasons, along with acclimation capacity, for the distribution patterns we have seen for these three sagebrushes.

Additional research demonstrated another aspect of big sagebrush's adaptability by showing that it was more drought-tolerant than dwarf sagebrush. Where these two overlapped, dwarf sagebrush always appeared on the moister sites. Interestingly enough, the kind of plasticity demonstrated by big sagebrush with regard to photosynthesis apparently doesn't hold for water loss by transpiration from the leaves. West found that big sagebrush from the subalpine zone showed essentially the same pattern of water loss even when transplanted to the warmer pinyon-juniper zone. Further, watered plants of big sagebrush in the latter zone showed a doubling of water loss compared to subalpine big sagebrush. Under these same conditions, dwarf sagebrush lost even more water than big sagebrush. When water was scarce, however, the

Dwarf Sagebrush

lower-elevation big sagebrush lost proportionately more water than the subalpine form. Under these stress conditions, black sagebrush lost more water than either the subalpine big sagebrush or dwarf sagebrush, which perhaps explains why it prefers the moister sites. Since dwarf sagebrush tends to have a high rate of water loss, this also may explain why it is restricted to the higher and colder elevations where evaporation would be reduced. Another interesting observation was that these three sagebrushes lose less water than any of the shrub or herb species they have as associates. There appears to be no special adaptation of their roots, and so we must speculate that they have some internal ability to resist drought to a greater degree.

Unfortunately, the facts are not really as simple as the story just told may indicate. H. B. Passey and V. K. Hugie—studying several species of sagebrush in Utah, Idaho, and Nevada—found that although black sagebrush is commonly associated with shallow, stony, and dry soils, all the soils on which it occurred were moderately deep, and only two were gravelly or stony. It was, in fact, found on deep silt loam as well. On the other hand, M. A. Fosberg and M. Hironaka studied the soil types associated with big and dwarf sagebrush in southern Idaho, and they found that dwarf sagebrush was restricted to soils with clay or bedrock within 33 centimeters of the surface. They concluded that the ability of dwarf sagebrush to tolerate the poor aeration characteristic of soils with clay near the surface was the primary factor controlling the distribution of the sagebrush in their study area.

These contradictions point up the extreme difficulty which plant ecologists face when they try to extrapolate results from one area to another. They can't always be certain which of the environmental factors is the most important or, for that matter, whether they have taken all the variables into account, including the possibility that physiologically different forms are involved.

In many of our desert and steppe shrub associations, the casual observer will note that there are relatively few young plants or none at all. Sometimes this is because it takes an unusual event, such as a range fire, to get the seedlings started, but their absence may indicate instead that the individual shrubs in the community are fairly long-lived. Passey and Hugie found that, depending on the site studied, the average age of the oldest big sagebrush plants varied from 27 to 100 years. The oldest plant observed was 120, as was determined by counting the annual rings of spring and summer wood in the main stem. Although Passey and Hugie did not count the annual rings for dwarf and black sagebrush, they assumed that the age of the oldest plants, based on nearby big sagebrush, was around 50 years. Our shrublands, then, in many instances are older than the adjacent forests at higher elevations or in moister locales. Most forest trees, of course, tend to be cut as soon as they are big enough to harvest, but generally, unless they are regarded as a nuisance, we leave the shrubs alone. Could one call this an example of the meek inheriting the earth?

West compared the three sagebrushes on various soil types and found that they were all the same in their ability to germinate—somewhat surprisingly, no significant differences could be found. Sometimes the distribution of plants is interpreted to be the result of certain limited circumstances controlling seed germination, but this is not the case with sagebrush. Later, in the seedling stage, big sagebrush appears to be more drought-resistant than dwarf or black sagebrush, with the latter being most sensitive to a reduced water supply. West found this to be true regardless of soil type. Since the surface layers of the soil will be heated by the sun, resistance to high temperatures becomes another critical survival factor. Black sagebrush seedlings were able to withstand higher temperatures on limestone soil than on sandstone soil. Dwarf sagebrush seedlings were slightly less resistant to high temperatures than were big sagebrush seedlings. Obviously, these contrasting responses of the seedlings to water and temperature may also help explain the differing distributions of these three sagebrushes.

Separating the dwarf and black sagebrushes in the field is sometimes difficult. There are no reliable and consistent differences in leaf shape between the two. In both, the lobes are sometimes shallow and sometimes very deep. If flowers are present, dwarf sagebrush is seen to have six to eleven flowers in each head, while black sagebrush has only three to five. The tiny bracts surrounding these flowers are mostly hairless in black sagebrush but are usually covered with fine, grayish hairs in dwarf sagebrush.

Although humans have some difficulty separating dwarf and black sagebrushes, some grazing animals apparently can do so with ease. Antelope prefer black sagebrush, and it is considered by some range experts to be a good browse plant for deer, goats, and sheep. The range sheep industry that evolved in the Great Basin was largely based on wintering on black sagebrush, which was regarded as hot feed that allowed sheep to survive on desert ranges in winter. Interestingly enough, black sagebrush contains a toxic principle which becomes manifest only if the sheep subsequently feed on horsebrush. One study carried out in Utah demonstrated that black sagebrush had high concentrations of phosphorus and vitamin A. Some years ago, Dick Holbo and I carried out a chemical study of these three sagebrushes, as well as other related species. This involved separating some of the chemical components by means of wood alcohol, notably the terpenoids considered under our big sagebrush discussion. Holbo found that each species and variety tested invariably had a characteristic terpenoid signature, and he could easily separate dwarf and black sagebrushes by this means. Although there was no way to know, we wondered whether grazing animals are somehow able to recognize the terpenoid signature of black sagebrushes.

A variation of this technique is used by some range botanists to identify sagebrush species. An extract made from the leaves is examined under ultraviolet illumination; the fluorescent color which appears can then be used to help determine the species or variety. Although many animals appear to be able to recognize the subtle variations in sagebrush quite easily, humans—with their relatively crude sensibilities—are just learning how to do the same, but only with the help of a lot of complex techniques.

Another interesting variable about which we have said nothing so far, mainly because of its complexity and because we don't really understand the way in which it affects the adaptability of certain species, has to do with the chromosomes found in *Artemisia*. The basic chromosome number in big sagebrush and related species is eighteen. Some years ago, George Ward of the University of Illinois studied the chromosomes of a number of sagebrush collections. He found that 33 percent of the big sagebrush examples, about 17 percent of the dwarf sagebrush examples, and over 80 percent of the black sagebrush studied had thirty-six chromosomes.

We call those races with multiple sets of the basic chromosome number polyploids; in many instances, they are indistinguishable from races with a single diploid set of chromosomes (in reality, a paired set of nine chromosomes each in sagebrush) and can be identified only under the microscope. Sometimes polyploid plants are larger and grow more vigorously than their diploid counterparts. One classic study of a diploid-polyploid species of *Sedum*, a type of succulent, showed that the polyploid race consisted of two types: one with two sets, called tetraploid, and one with three sets, called hexaploid. The most vigorous of the three races and the one with the smallest distribution was the hexaploid race. The botanist studying the succulent concluded that the hexaploid race had only recently evolved and was destined to replace the other two races, except where special circumstances might favor them—the diploid race, for example, grew better than the other two when the tannic acid content in the soil was higher.

Possibly some similar pattern of evolution is occurring in sagebrush, but we can only speculate about this, since no one has studied any different adaptability or physiology which might be associated with polyploid forms of sagebrush. In any event, this is another way in which evolution fine-tunes the physiology of individual species to adjust to the environment. Cellular and physiological adjustments of this kind are constantly going on in nature, and we are only now beginning to appreciate both the beauty and the complexity of the process. This could be considered a kind of microevolution, and even though we don't live long enough to see new species or genera evolve we can at least learn a great deal by watching the generation of new races or varieties.

The generic name for sagebrush comes from Artemis, the Greek goddess, Apollo's sister and Zeus' daughter. She was reported by a Greek philosopher of the second century to have delivered a medicinal plant to Chiron the Centaur, who named it artemisia. There are over a hundred species of *Artemisia* distributed throughout the northern hemisphere and into South America; about nine woody species occur in or very near the Great Basin. Many *Artemisia* species are not woody, but the woody forms which we have constitute an important and conspicuous portion of the flora of the Great Basin.

Silver Sagebrush
Artemisia cana

SILVER SAGEBRUSH, sometimes called hoary sagebrush, perhaps would be most easily mistaken at a distance for big sagebrush. The species name *cana*, meaning white or gray, refers to the distinctive silky, silvery appearance of the leaves. They generally have no lobes at the tip, as do those of big sagebrush, and they are long and narrow—typically about 5 centimeters long and a little less than 1 centimeter wide. The shrubs vary in height from about .5 to 1 meter, at their largest being only about 1.5 meters tall. The twigs are densely pubescent, and older branches have a brown, fibrous bark. A very distinctive feature of this species is its ability to root-sprout; this is particularly noticeable after a fire. Some roots will give rise to buds on their surface, which are capable of growing upward and developing the anatomy of stems rather than roots.

In addition, silver sagebrush is unique in its ability to produce elongate, underground stems which help the shrub spread. Unlike big sagebrush, which bears its flowers on narrow stems with few leaves above the main body of the plant, silver sagebrush has its flowers borne on densely leaved branches. Flowering occurs during August and September.

The subspecies of silver sagebrush in the Great Basin is found commonly at high elevations, along streams or where some moisture is present well into the growing season. Another subspecies of silver sagebrush, A. c. subspecies *bolanderi*, occurs in the Sierra Nevada, but it is separated by a gap of about 170 miles from that which occurs from central Nevada on to the east. In 1960, in his intensive study of the shrubby *Artemisias*, A. A. Beetle recognized three varieties of silver sagebrush and considered it second only to big sagebrush in total acreage occupied in the eleven western states. Big sagebrush was estimated to cover some 226,374 square miles, while silver sagebrush occupied some 53,221 square miles. Within the Great Basin proper, however, it is much less common than either black or dwarf sagebrush. In terms of usefulness for grazing, silver sagebrush is a little more palatable than big sagebrush but not nearly as preferred as black sagebrush.

Silver sagebrush has the distinction of being the first of the shrubby western *Artemisias* to be described botanically. Frederick Pursh published a description of it in 1814, from a collection made ten years earlier along the bluffs of the Missouri River by Meriwether Lewis.

Bud Sagebrush
Artemisia spinescens

BUD SAGEBRUSH is a nearly ubiquitous companion of shadscale throughout the Great Basin. It should be no surprise, then, that it is adapted to much more arid conditions than are other shrubby sagebrush species. Its wide range accounts, perhaps, for some of its other common names—button brush, budsage, spiny sagebrush, and spring sagebrush. Like the other sagebrushes, it has a characteristic pungent, aromatic odor. However, it differs considerably in form from our other shrubby *Artemisias* and, in fact, is not included in the same subdivision of the genus. The immediate relatives of bud sagebrush are considered to be some of the herbaceous weedy forms which show no woody characteristics.

Bud sagebrush is the only spiny sagebrush in the West, and as such its spines have an origin different from that of the spines of our other desert shrubs. Each spine develops from an axis that supports several flowering heads. After the heads have flowered and fruited, they break off and leave behind sharp-pointed, woody stalks up to several centimeters long. This is but another example of those opportunistic evolutionary processes which enable plants to adapt expeditiously to changing environmental conditions in a way that, to the uninitiated, seems so deliberately purposeful. Spines, characteristically present in many arid-land plants, are obvious devices that help protect plants from excessive grazing pressure. The succulent members of the spurge family in the deserts of Africa have several types of thorns, one of which develops in a fashion similar to that of bud sagebrush: in this instance, the entire flower stalk frequently aborts and forms spines instead of flowers. Some forms, such as cacti, have the leaves and bud scales very much modified to form spines. In many cases, as in spiny hopsage, ordinary

Bud Sagebrush

vegetative shoots may be modified to form spines. Interestingly enough, in a few plants, spines will develop under conditions of low humidity, but fail to form under moister conditions.

The most complete life history study ever undertaken on bud sagebrush was conducted by Benjamin W. Wood for his master's degree at Brigham Young University. Among other things, Wood found that bud sagebrush stems show the same lobing habit as big sagebrush stems, for basically the same reason. That is, the cambium becomes inactive in segments around the stem, and as a consequence continuous rings of wood will not be formed in mature stems. Eventually, a single stem may separate into two or more stems when the center decays. Also, as in big sagebrush, a layer of cork is formed each year between the new annual ring of wood and last year's ring, with the result that the upward movement of water is restricted to the current year's new wood.

Like many other plants adapted to the desert, bud sagebrush has a long taproot in deep soils lacking a hardpan, as well as numerous fine, horizontal roots confined to the top 15 to 50 centimeters of soil. The roots also develop a layer of cork separating each successive annual layer of wood. In an interesting observation made after a hard desert rainstorm that lasted about twenty minutes, Benjamin Wood found that moisture had penetrated to a depth of 10 centimeters beneath bud sagebrush but only to a depth of 4 centimeters beneath shadscale. Similar studies in Australia have shown that various desert shrubs differ in their ability to capture moisture from brief showers. The fine, horizontal roots of many shrubs are adaptations to the rather shallow penetration of the soil by such precipitation.

Bud sagebrush leaves are said to be palmately divided into several narrow lobes, in other words, like the fingers of a hand. Each of these lobes is in turn divided into three lobes. The leaves are covered with a matted, woolly pubescence. They begin to bud out very early in the Great Basin, frequently in February, when most other shrubs are still dormant, and tend to dry up and fall by midsummer. Unlike most other plants with deciduous leaves, in which an abscission layer forms at the juncture of the stem and the leaf stalk that aids in leaf fall, bud sagebrush develops an abscission layer higher up on the leaf stalk, above the expanded basal portion. The result is that, after the leaves fall, the basal portion remains behind as a protection for the next year's buds—only after these buds have expanded will the rest of the leaf stalk become detached. The young stems are white-pubescent, while the older stems are characterized by a brownish, fibrous bark.

Unlike big sagebrush, bud sagebrush's small clusters of greenish yellow flowers or "heads" have strap-shaped or ray flowers in each head, like those of the daisy, albeit yellower in color and much smaller. Bud sagebrush begins to flower, depending on the weather and the site, from March until June. The flowers in the center of each head remain sterile; only the ray flowers produce the little, hard, fruiting bodies or achenes.

Wood, studying the seedling development of bud sagebrush, found that if the seedlings were allowed to dry out during the first twenty to thirty days they would die. In the first year of growth, the stems were only about a centimeter high, while the roots grew 12 to 20 centimeters deep. As in many other desert shrub seedlings, the theme seems to be to use most of the photosynthetic energy during the first several years to establish a deep and efficient root system. On the other hand, many desert annuals have only shallow and feeble root systems, and they flower and fruit and then die before the hottest part of the summer begins. This kind of avoidance strategy is one of the most successful forms of adaptation among both plants and animals in a desert environment.

Wood found that bud expansion began in March or early April; leaf development and twig elongation followed and were completed by June. During the season he studied these plants, all the leaves had turned brown and died by the first week of July. However, a late summer interval of precipitation caused a second period of bud expansion to occur, and new, small leaves were produced. Blooming began the last week of April and continued until the last of May. Bud sagebrush, despite its small size, has a much more branched root system than any of its companion shrub species. Wood found horizontal branch roots as long as 2.1 meters and taproots as deep as 1.8 meters.

How long do desert shrubs live? For most shrubs there are simply no good figures available, but in general, judging from the few studies that have been made, we can say longer than might be expected under such harsh conditions. Wood tried to count annual rings in bud sagebrush, but this technique did not work very well because of the lobing pattern of growth in the stems. However, in 1964, by remapping a protected plot first studied in 1935, he found that about 24 percent of the shrubs were still alive; these plants, at least, must have been more than twenty-nine years old.

Bud sagebrush is regarded as a worthwhile forage plant in areas where little other forage can be found. Sheep appear to consider it most palatable early in the spring, when the buds begin to show signs of breaking dormancy. At this time, pulling on the twigs will easily separate the bark from the last season's growth—sheepherders refer to this phenomenon as slipping. Since bud sagebrush is one of the first desert shrubs to show signs of growth toward the end of winter, it was once regarded as an important shrub for the range sheep industry. For cattle and horses, it is regarded as poor to fairly useless forage. However, for a variety of wildlife species, including deer, it is a very important plant in the food chain.

Bud sagebrush thrives on a variety of soils, from clay loam and loam to sandy loam and sand. Wood found that the soils ranged from mildly alkaline to strongly alkaline, although the salinity varied from none to only a small amount. The shrub occurs from southeastern Oregon to western Wyoming south through most of Utah and Nevada to southern California, northern Arizona, and western Colorado. It is found in virtually all the desert communities within the Great Basin. Its associates include not only shadscale but winterfat, galleta grass, Bailey's greasewood, spiny hopsage, green molly, and even big sagebrush.

Bud sagebrush gets its common name from its budlike clusters of flowers and leaves; the species name obviously comes from its spiny nature. Unlike many of our other sagebrushes, bud sagebrush is very uniform over its entire range—there are no recognized varieties or subspecies. Why this is the case has no really adequate explanation. The family to which bud sagebrush belongs is the Asteraceae or aster family, some of the characteristics of which are discussed in our treatment of the brickellbush and littleleaf horsebrush.

Big Sagebrush
Artemisia tridentata

It never rains here, and the dew never falls. No flowers grow here, and no green thing gladdens the eye. The birds that fly over the land carry their provisions with them. Only the crow and the raven tarry with us. Our city lies in the midst of a desert of the purest, most unadulterated and uncompromising sand, in which nothing but that fag-end of vegetable creation, "sagebrush," ventures to grow. —Mark Twain, LETTERS

PROBABLY NO OTHER plant is so evocative of the Great Basin as big sagebrush, at least to those of us who know and love the area. And we would hardly agree that the sagebrush is the "fag-end of vegetable creation"! Mark Twain's colorful condemnation is about as accurate as his description of iron doors flown as kites by street urchins with the aid of our frequent Washoe zephyrs.

A much more characteristic attitude is typified by one of the patriarchs of the Biology Department at the University of Nevada, who related that during his service in World War II he grew so homesick that he asked his mother to send him a twig of big sagebrush in order to experience the Great Basin once again. No one acquainted with the penetrating pungent odor of big sagebrush—also known as blue sage or black sage or sometimes simply as sage—after a summer shower can forget it or fail to visualize the vast expanses carpeted with this gray-green shrub, which seems to occur almost everywhere in the Basin. On reflection, that is too general a statement, for in reality big sagebrush does not occur everywhere here. It is not really a desert plant but is typically found in so-called steppe areas, with rainfall of from about 18 to about 40 centimeters a year. It is dominant in the foothills of our mountain ranges and continues on up into the pinyon-juniper woodlands. Sometimes big sagebrush can be found in the drier valley bottoms, but only along gullies or in other spots associated with water.

Although it is incorrect to say that this shrub "searches" for water, it sometimes seems so, for sagebrush has a stout taproot which grows to a depth of between 1 and 4 meters and frequently penetrates to the capillary zone just above the water table. In addition, many radially spreading lateral roots near the soil surface insure that the moisture from a light rainfall will also be readily absorbed. R. D. Tabler, in a study at an elevation of 9,500 feet in Wyoming, showed that big sagebrush had 62 percent of its root length concentrated within the upper 60 centimeters of soil. Within this same distance, the taproot tapered to a diameter of only 2 to 3 millimeters. Some roots, however, penetrated to a depth of nearly 2 meters, while some spread outward from the main axis to a distance of 1.5 meters.

Because of its ability to absorb moisture from the soil, big sagebrush is considered by range scientists to be a phreatophyte, a plant which robs the soil of water. However, one experiment by Joseph Robertson, a range ecologist for the College of Agriculture at the University of Nevada at Reno, showed that big sagebrush is not all that effective when it comes to competition with drought-resistant grass. Big sagebrush suffered considerably in competition with crested wheatgrass when the roots of both were restricted by a hardpan about a third of a meter beneath the soil surface. Subsequently, big sagebrush grown in containers with wheatgrass died when water was restricted, but the grass survived. Apparently, the deep root system of big sagebrush is one of its most effective devices for coping with a severe water shortage.

Big sagebrush over a meter tall is considered a good indicator of arable land, for it prefers deep, moderately basic soils. The present amount of sagebrush within the Great Basin appears to be greater than was the case before the area was settled. In fact, it is now abundant in parts of the Pacific Northwest with rainfall of 50 centimeters a year that ought to be dominated by grasses, the so-called Palouse prairie. The success of big sagebrush in these moister areas is due primarily to the fact that cattle find it unpalatable, and excessive grazing has eliminated the natural grass competitors and allowed this shrub to spread—in many cases almost unopposed. Horses will seldom eat more than a few sagebrush flower heads. Sheep and goats find it somewhat more palatable, especially in winter, when little else may be available.

Although no one has determined what it is about big sagebrush that cattle find unappetizing, one good guess is that this probably is due to a group of compounds, known as terpenoids, found within the leaves. Certain other sagebrush species which are more palatable to cattle have a lower concentration and a reduced variety of these terpenoids. Whatever the explanation, if it were not for protective compounds, there would undoubtedly be much less big sagebrush in the West, for its leaves equal alfalfa meal in protein content and have more carbohydrates and twelve times more fat. Its leaves and flowers constitute most of the diet of the sage grouse and are also a primary food source for antelope and deer. Recent studies have shown that deer apparently avoid the influence of the volatile compounds in sagebrush by "belching" them as they chew their cuds. Why can't cattle do the same? Although it furnishes only a minor percentage of the diet for such animals as the jackrabbit and ground squirrel, big sagebrush is invaluable in providing needed cover for a host of desert animals.

According to one estimate, big sagebrush is the most abundant shrub in North America. Its natural range extends from western Nebraska to Montana and British Columbia and south to eastern California, lower California, and northern New Mexico. In Oregon it ascends to 3,000 feet, while in Colorado it reaches timberline. In Nevada, the creosote bush replaces big sagebrush as a dominant shrub in the area south of a line extending east roughly from the area of Tonopah. Oddly enough, there is also one report from forty years ago that big sagebrush had become well established in an area near Concord, Massachusetts. Certainly, there is no question that, of all the species of sagebrush, big sagebrush is best adapted to a wide variety of habitats.

Because of its aggressiveness, big sagebrush has been the object of a great many studies by range scientists interested in finding ways of eliminating it in favor of grasses, both native and introduced. Joseph Robertson, in one of his many studies on this plant, characterizes big sagebrush as "definitely undesirable (in dense stands). It is relatively unpalatable to sheep and cattle, and it uses moisture and nutrients that should be producing good forage. . . . Sagebrush prevents grazing of grasses hidden under its woody stems and crown. It hampers movement of livestock, especially sheep. The brush snags wool from fleeces. It causes lambs and calves to stray and become lost, and heavy brush makes conditions ideal for predators such as coyotes." This same source, however, takes a balanced view of the value of big sagebrush in the natural environment by pointing out that "control on ranges used by big game in the winter is not advocated where sagebrush is an important source of browse. A balanced mixture of broad-leaved herbs, grasses, and shrubs is considered desirable for most wildlife of these sagebrush ranges."

Although it is not really a desert plant, big sagebrush is considered by many botanists to be a xerophyte (from the Greek *xeros*, which means dry). The leaves at all stages of development are clothed with a dense layer of short hairs, which are dead and hollow at maturity. These hairs reflect some of the light and are responsible, along with the chlorophyll in the leaves' interior, for their greenish gray, silvery appearance. In the fashion already described in the introduction to this book, hairs on leaves help reduce water loss in plants, although no one has determined to what extent they might help sagebrush. One interesting thesis postulates that, where there is a sufficient supply of subterranean water, plants able to grow roots down to this level will need only periodic protection from the drying effects of wind—in such instances, the covering of hairs on the leaf surface provides this protection. On the other hand, according to this same theory, where the water supply is limited and continued protection is required, a waxy or varnishlike cuticle works best, as is the case with the leaves of the creosote bush.

In addition to the dead hairs on the surface that help retard water loss, big sagebrush leaves also have living hairs that are glandular in nature and appear to secrete the characteristic ethereal oils, some of which account for this shrub's unique odor. It has been suggested that the abundant production of volatile oils which blanket the surface of the leaf may be effective in reducing water loss.

The leaves of big sagebrush are quite variable, in both size and shape. In general, they are about 2 or 3 centimeters long and about two-thirds of a centimeter wide, in the shape of a wedge attached at the narrow end to the stem. Some leaves, however, may be as long as 6 centimeters and well over 1 centimeter wide. Typically, they have three shallow lobes at the tip—the specific name *tridentata* means three-toothed—although leaves with more or fewer lobes can be found on almost any plant. Leaf size and the amount of lobing are good indications of the available moisture at the time the leaves were forming—those formed during the moist spring will be larger and have more lobes, while those formed later will be smaller and perhaps lack any

Big Sagebrush

lobes. Quite frequently, the leaves will be borne in dense, rosettelike clusters at the stem tips as well as individually along the stems.

Whatever their shape or arrangement, the primary function of leaves is photosynthesis or food manufacture. All other things being equal, the more efficiently a plant carries out this process, the more competitive it is. One study, carried out by E. J. DePuit and M. M. Caldwell, determined that big sagebrush had the highest rates of photosynthesis in late May and early June; then a decline began, with the lowest rate occurring in August. Early in the season, the rate was controlled by the temperature and the amount of light available. Later, however, the dry months reduced the rate of photosynthesis by causing the stomates or leaf pores to close early in the morning and remain closed for the rest of the day. An additional decline was apparently caused by the tendency of leaves, as they age, to become less effective at carrying on food manufacture.

Some adaptation to the seasons occurs, however, in several ways. Early in the spring, the optimum temperature for photosynthesis is 15 degrees C., while later it increases to 20 degrees C. In other words, big sagebrush is better able, compared to certain other shrubs later in the season, to accommodate to higher temperatures, insofar as food manufacture is concerned. Another adaptation occurs with the ability of the plant to produce smaller, denser leaves during dry periods. The large leaves produced during the spring are poorly adapted for midsummer conditions and typically die and drop from the plant by late June. Most of the stem growth takes place during the first two weeks of June, when the greatest amount of food is available from photosynthesis.

Many mature leaves remain on big sagebrush throughout the winter, thus making it technically an evergreen. Most other shrubs have winter buds and various types of bud scales to protect them from the rigors of cold weather, but big sagebrush gets by without winter buds. The leaves simply stop their growth at the beginning of the cold season and appear to go into a kind of suspended animation until spring revives them, whereupon they start to grow again. These overwintering leaves are lost early in the next growing season, however. Because of this, it is probably better to call big sagebrush a wintergreen rather than an evergreen plant.

While many leaves on big sagebrush are borne in a more or less horizontal plane, as are the leaves of most plants, there are also many others borne in every other conceivable plane. It is thus not surprising to find that sagebrush leaves have stomates in about equal numbers on both surfaces. In most plants with horizontal leaves, the stomates are located only on the lower surface.

Another possible way in which big sagebrush has become adapted to its environment is through a phenomenon known as allelopathy. Many plants, such as the creosote bush and purple sage, are able to produce a compound in their leaves which will inhibit the growth of other plants. As a consequence, when the leaves fall, the leaf litter beneath the plant decreases the ability of other seeds to germinate and grow. Whether sagebrush is capable of this kind of competition has been the subject of a few, somewhat inconclusive studies. An experiment conducted by E. F. Schlatterer and E. W. Tisdale showed that a water extract of sagebrush litter inhibited the germination of several kinds of grasses in the laboratory. However, the amount of inhibition obtained was so small that they doubted that sagebrush depended much on allelopathy as a competitive device under field conditions. And, to further complicate matters, four weeks after germination the growth of grass seedlings was stimulated by this same litter. At any rate, every observer of our natural scene knows that many of our native wildflowers find shelter beneath the branches of sagebrush, so that it appears to be a rather benign competitor, at least insofar as allelopathy is concerned.

In 1938, R. A. Diettert of Montana State University published a very complete study of the anatomy and development of big sagebrush. It appears, oddly enough, that the type of internal leaf anatomy found in big sagebrush and many other xerophytes is actually conducive to greater water loss rather than less, as might be expected. The peculiar features of the sagebrush leaf seem to be primarily an adaptation to high light intensity. The shrub copes with drought conditions by shedding some of its leaves and by means of a relatively high concentration of salts in the cell sap. Roughly, the more salts or other soluble materials (such as sugar) present, the more negative is the cell potential, and the more resistance there is to the loss of water by that cell. For most plants adapted to moister atmospheric conditions, the cell sap averages around minus ten atmospheres or less (one atmosphere being the pressure exerted at sea level by the air), but for big sagebrush the figure may be as low as minus seventy-three atmospheres. This high sap concentration undoubtedly aids big sagebrush in adapting to relatively dry conditions. However, many shrubs growing in drier and more saline parts of the Great Basin have considerably more negative cell potentials.

Under the best conditions, big sagebrush may reach a height of 4.5 meters and have stem diameters of over 10 centimeters. The bark on older stems is gray-brown and shreds readily from the plant. A cross section of an older stem will show that the rings of wood laid down each year are not of uniform width around the center of the stem. This characteristic growth pattern gives rise to the notable eccentricity of sagebrush stems, with their twisting, tortuous appearance. Probably this is also somehow adaptive for the sagebrush. Sometimes the bark shreds sufficiently from the stem to expose the cambium, which then dries out and dies. This also adds to the irregularity of the stem.

One of the striking and unique features of the internal anatomy of the big sagebrush stem is the presence of a cork layer beneath each annual ring of woody tissue. In most other woody plants, when cork forms, it develops only outside the woody and food-conducting tissues and composes the outermost layer of the bark. Characteristically, when fully mature, cork is impervious to water and also to gases. Although sagebrush also has cork in its bark, additional cork is developed within the woody, water-conducting tissues. Each year, as the growing season begins, a layer of cork is initiated which lies between the new wood ring and that of the previous year. By early August, the cork layer is two or three cells thick; it is fully formed, and the wood enclosed by it no longer contains any living cells. Consequently, only the most recently formed annual ring of wood remains capable of transporting water to the upper part of the plant. Although we can't be certain, it seems probable that these cork layers within the wood protect sagebrush from excessive drying, especially when many of the annual shoots die back. Additionally, the layers may protect the stem from invasion by fungi or bacteria or damage from animals. Nevertheless, one thing is certain: these anomalous layers serve to restrict water transport to a relatively narrow zone of wood, compared to other woody plants, where at least several outer annual rings of wood are functional in water transport. Other species of sagebrush are known to develop these cork layers within the wood—see the discussion of bud sagebrush, for example—but they are invariably inhabitants of arid environments.

Still another way in which big sagebrush apparently adapts to drier summertime conditions was discovered by Stephen Dina in his studies in Red Butte Canyon, Utah. Dina found that sagebrush subjected to arid conditions underwent a significant decrease in the amount of nitrogen stored in the leaves, while the amount of that element stored in the roots and stems showed a definite increase. Interestingly enough, these changes are similar to those found in other plants whose leaves are in the process of growing old and beginning to die. The changes are believed to signify a breakdown of protein within the leaves and an increased protein synthesis within the stems and roots. Of course, after the leaves die, some of the nitrogen which would otherwise be lost is conserved. We might ask, however, what stimulates these changes within the plant. We now believe that the trigger is the increased water stress which reduces the supply of a hormone, known as cytokinin, needed for protein synthesis. This hormone, manufactured in the roots, is transported upward in the stems. In some unknown way, the transport of this hormone is interfered with, and the result is a decline in protein synthesis and aging of the leaf.

Actually, the process of aging in sagebrush leaves is probably still more complicated than this. Although no one has investigated the situation in big sagebrush, it has been discovered that in many other plants a specific dormancy factor, known as abscisic acid, is produced when they are stressed by drought. This dormancy factor slows growth and protein synthesis and apparently also causes the stomates to close. This latter effect is just the opposite of that of cytokinin, which stimulates the stomates to open. Plants which are periodically stressed by drought have a high level of abscisic acid and thus remain stunted. This, of course, can be viewed as one more way in which desert plants are able to adapt. Also, along with the decline in the protein content in the leaves goes an increase in sugar content; the stems and roots experience a similar increase. A higher sugar concentration, as has been pointed out, probably also helps the plant resist further drying.

Nor have we exhausted the possibilities, for yet another way in which sagebrush can adapt was shown in a study by H. A. Mooney and Marda West in the White Mountains of California. They grew desert plants of five different species from seeds and acclimated the seedlings for several weeks to either a desert or a subalpine environment. They found that two species, the sagebrush and the fern bush, showed the most plastic responses. Those seedlings acclimated to desert conditions carried on photosynthesis best at

25 degrees C., while seedlings grown under subalpine conditions were more productive to 20 degrees C. This may be another reason why sagebrush is so widespread in the West.

Flowering in big sagebrush occurs from August until well into October and may be stopped only by the onset of cold weather. The flowering branches are usually thinner and quite erect, and—although the leaves at the base are the typical three-toothed form—those further up are narrower and lack the three apical lobes. Since big sagebrush is a member of the aster family, its flowers are really clusters of three to twelve tiny flowers grouped into a head and enclosed by small bracts. The heads individually lack stalks and are only 3 to 4 millimeters tall; typically they are clustered, with three to seven in each group. They are relatively inconspicuous and somewhat yellow-green at maturity, in keeping with the fact that they depend only on the wind for pollination. The development of the individual flowers in each head is unusual in that the petals are interlocked and fused together—they open eventually only because the stamens enlarge and force their way upward. As with most wind-pollinated plants, pollen is produced in abundance, and unfortunately it is the bane of many hay fever sufferers.

After pollination and fertilization, each flower produces a single, usually reddish brown seed enclosed in a woody coat, similar in structure to a sunflower "seed" with its hull, only very much smaller. Duwayne Goodwin, who conducted a very thorough study of the life history of big sagebrush, found that while the seeds varied somewhat in size they were a little over a millimeter long and about seven-tenths of a millimeter wide, making them smaller than the seeds of many of their associates. Because the yellowish hull encloses air, they will float on water, which may explain why big sagebrush is so common along watercourses in the desert. The small size of the seeds also favors their being transported in the fur of domesticated and wild animals. Speculation has it that the sagebrush colony near Concord, Massachusetts, was the result of using wool combings from western sheep as fertilizer.

As is true of most small-seeded plants, big sagebrush produces a prodigious quantity of seeds. Goodwin estimated that an average big sagebrush with a crown a meter in diameter produced about 450 flowering branches and at least 350,000 seeds! Really vigorous plants in a good year probably produce over a million seeds. Certainly this observation helps explain why big sagebrush is such a widespread and successful shrub. Dispersal of the seeds begins in October and continues into January. Once germinated, the seedlings grow rapidly, under good conditions, and may begin to flower within four years.

Quite often, desert plants are characterized by seeds which fail to germinate unless they are exposed to cold, have their tough coat scratched open, or undergo some kind of afterripening period. This does not appear to be the case with big sagebrush, however. Usually its seeds will germinate immediately after being harvested, although some exposure to cold does appear to improve the germination percentage. Interestingly enough, Goodwin found that exposing the seeds to light frequently resulted in increased germination. This phenomenon is well known for the seeds of certain other plants, such as the saguaro cactus and lettuce. Lettuce seeds will germinate if exposed to red light, but germination will be inhibited if they are exposed to far-red light of a slightly longer wavelength; a pigment known as phytochrome, which occurs universally in plants, is responsible for this light-sensitive reaction. Very probably, this same phenomenon will be found to occur in big sagebrush.

One of the natural processes that big sagebrush does not appear to be well adapted to is fire. It seldom root-sprouts after a fire, depending primarily on seeds buried in the soil or carried in by animals to recolonize a burnt area. However, since the seedlings compete poorly with grasses, reestablishment is frequently delayed indefinitely. Deliberate burning has been used to control undesirable stands of sagebrush, and this can certainly be considered the most "natural" method of control. Unfortunately, it can also be the most dangerous method if not controlled carefully. Particularly for the drier plant associations in the country, fire is a very natural and periodic phenomenon, and many plants have over the millennia evolved a variety of ways of coping with it, but big sagebrush is not among them.

Within the Great Basin, big sagebrush commonly grows in association with rabbitbrush, green ephedra, spiny hopsage, bitterbrush, horsebrush, and the plateau gooseberry. When not overgrazed, the sagebrush zone tends to be dominated, as discussed earlier, by grass, particularly bunchgrass. One study found that a cemetery protected from grazing at Mona, Utah, had bunchgrass as 92 percent of the total cover, while big sagebrush made up only 1 percent. Outside the cemetery, big sagebrush dominated completely. Within the sagebrush zone of Utah, junipers are invading the overgrazed sites. Dwight Billings, who first clearly defined the plant associations of the northern Nevada desert, discovered that the pinyon pine invaded overgrazed areas in the western portion of the Great Basin. Even without overgrazing, however, some areas tend to be dominated by big sagebrush. Christensen and Johnson believed that this might be the result of the timing of precipitation—precipitation coming primarily in the winter favors sagebrush rather than grass, probably because the penetrating roots of sagebrush are able to bring up water from the deeper soil layers during the dry summer.

Although the European and Asiatic relatives of sagebrush—identified under the general name of wormwood—were used for a variety of medicinal purposes, only our native Americans seem to have made any significant use of big sagebrush. A tea made from its leaves was considered effective as a tonic, as an antiseptic for wounds, as a remedy for colds, sore eyes, and diarrhea, and as a way of warding off ticks. One of its more unusual uses was as a hair tonic. Also, the seeds were eaten raw or pounded into meal. About the only use the settlers could devise for sagebrush was to burn it. The wood ignites easily, burns with an intense heat, and has been utilized in mine smelters. Recently, with the worldwide shortage in petrochemicals, interest in attempting to extract useful compounds from sagebrush has revived. So far, however, all these efforts are very limited experimental ones.

Those botanists who have studied big sagebrush in some detail are able to distinguish several varieties, which, while they differ only slightly in appearance, are nevertheless consistent in their distribution patterns and habitat. Strictly speaking, in the case of big sagebrush, these are referred to as subspecies rather than varieties. Don't ask what the difference is between a variety and a subspecies—the botanists can't agree among themselves! We should remember that we invented the pigeonholes of species, variety, etc., but that nature goes blithely on, serenely and forever evolving new forms in its own way, which may or may not fit our desire to classify everything.

At any rate, there is an old dictum that every visible difference in plants has an underlying physiological basis. That this is the case with the subspecies of big sagebrush was illuminated by Alma Windward, who worked on the species in Idaho for his doctorate. He found, for example, that *A. tridentata* subspecies *wyomingensis* grew in the driest habitats along with needle and thread grass. The moister sites were occupied by the subspecies *vaseyana*, sometimes with needle and thread grass, at other times with blue bunch wheatgrass, Idaho fescue, or mountain bromegrass. Clear evidence that the difference between these subspecies was more than superficial was shown by their difference in palatability to grazing animals: Windward found signs of grazing on the subspecies *wyomingensis* but none on the subspecies *tridentata* growing nearby. It does appear that the subtle differences between the various subspecies of sagebrush must be studied by ecologists if they are to adequately understand the climatic and soil factors controlling the distribution of sagebrush over the vast areas of the West.

Littleleaf Brickellbush
Brickellia microphylla

THERE ARE, perhaps, one hundred species of *Brickellia* distributed through the warmer regions of the United States south through Mexico and Central America to South America. Some are herbaceous rather than shrubby, and the only shrubby ones distributed well within the Great Basin are the littleleaf and the California brickellbushes. The oblongleaf brickellbush occurs on the western edge of the Basin in Washoe County; on the eastern edge in western Utah, it is found as far north as Salt Lake City. All our brickellias are shrubs of rocky canyons and washes, frequently growing out of rock crevices in very dry locales within our desert ranges. Their inconspicuous aspect frequently causes them to be overlooked, since only a few usually occur at any one spot. The littleleaf brickellbush characteristically occurs on the driest sites.

A typical littleleaf brickellbush consists of a few erect, branched stems with elliptical leaves between 1 and 2 centimeters long; sometimes there are a few blunt teeth along the leaf edge. During late summer and fall, the stems will be topped by several clusters of flowering heads, each about a centimeter tall and each bearing up to two dozen tiny, cream-colored flowers. The leaves and stems are covered with an obvious dense but short pubescence which is also glandular—a term used to signify a special type of hair which, depending on the species involved, is capable of secreting a variety of substances, sometimes volatile oils and sometimes sticky materials that effectively coat the surface of the plant. This mucilaginous secretion frequently helps protect the plant against insect predators. In fact, the similar but somewhat larger sticky glands of the carnivorous sundews (*Drosera* species) trap insects so effectively that the plant has evolved an enzyme system able to digest virtually all the soft parts of insects and thus make use of the

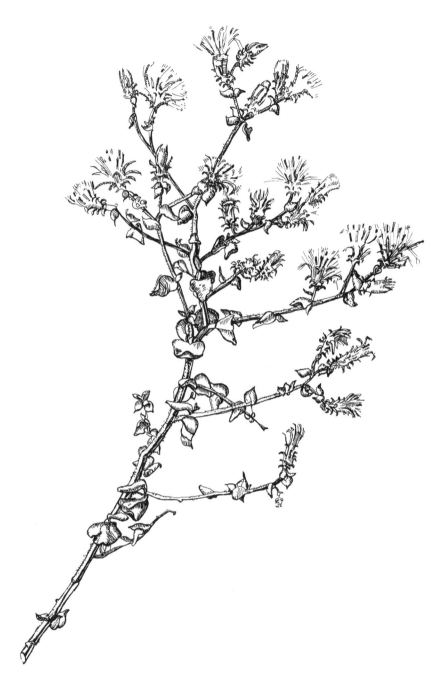

Littleleaf Brickellbush

nutrients they contain. Glandular hairs are easily recognized under a hand lens by their swollen, bulbous tips.

Like the rabbitbrushes and sagebrushes, brickellbush is a member of the aster or composite family. And consequently, like sagebrushes, each head is really a cluster of small flowers surrounded by numerous tiny bracts. Shortly after blooming, the flowers are followed by narrow fruits about 5 millimeters long. Each fruit, known as an achene, is really a dry, woody hull enclosing a single seed—all the members of the aster family produce this type of fruit. Frequently, as in the brickellbush, the achene is topped with fine, long bristles; these are assumed to be all that remains of the sepals we normally expect to find beneath the petals. In *Brickellia*, the petals are fused into a narrow cylinder which is also perched on top of the achene. The achene really represents the ovary of the pistil.

This kind of arrangement of flower parts is, as we discussed briefly under the description of the American dogwood, known as inferior, since the ovary is below everything else. Inferior flowers are regarded as advanced on the evolutionary scale, and the combination of inferior flowers, fused petals, very much reduced or even absent sepals, and a number of small flowers crowded together in heads causes many botanists to consider the aster family as being among the most advanced of flowering plant families. Evolution in plants involves not only increasing complexity at some levels but also a reduction or even a loss of parts at other levels. It has taken botanists some time to realize that the simplest flowers, so far as numbers of parts and construction are concerned, are usually fairly advanced rather than primitive.

Littleleaf brickellbush ranges from California to Oregon and Idaho and south through the Great Basin to Utah. It is found from elevations of 3,000 to 8,000 feet. The genus name honors J. Brickell, a Georgia botanist. The species name, as might be guessed, means little leaf.

Oblongleaf brickellbush, *B. oblongifolia* variety *linifolia*, is easily distinguished. It has larger flower heads, up to 2 centimeters tall and each borne singly rather than in clusters. The individual flowers are yellow to purplish, and the plants appear lighter because of a white pubescence. In addition to its presence on the eastern and western edges of the Great Basin, oblongleaf brickellbush extends into southern Nevada and northern Arizona and east to Colorado and New Mexico.

California brickellbush, *B. californica*, is also easy to separate. Although its flower heads cluster, the small bracts surrounding them are lance-shaped and straight—in the littleleaf brickellbush, these bracts curl downward at the tips. *B. californica* is very widespread throughout the western states, extending east to Texas and Oklahoma and north to Idaho, Oregon, and northern Colorado.

Another species, known as the tasselflower, *B. grandiflora*, is usually considered a herbaceous form but is occasionally slightly woody and may be mistaken for a shrub. Its flower heads are clustered at the ends of the branches and generally hang downward, accounting for its common name. The triangular leaves are not blunt, as in the California brickellbush, but have longer, pointed tips. The tasselflower frequently has much larger leaves, which on occasion attain a length of 11 centimeters, compared to a maximum of about 4 centimeters for the former species. It was once introduced into cultivation, though it does not appear to be grown commercially now, except in nurseries specializing in native plants. It ranges from California to Washington, Colorado, and Arizona, including the Great Basin.

None of the brickellbushes are considered significant range plants—the standard *Range Plant Handbook* of the Forest Service ignores all the members of this genus. There are records of the use of *Brickellia* by Gambel's quail and possibly other birds. The literature evinces very little use of brickellbush by the Indians. However, it appears to have been sparingly used as a stomach or headache remedy.

Rubber Rabbitbrush
Chrysothamnus nauseosus

In late summer, rabbitbrush, a gorgeous plant despite its unflattering specific name which means "nauseating," lines the highways and clusters in clumps between the sagebrush with a striking combination of golden flowers and silver-green foliage. —Elna Bakker, AN ISLAND CALLED CALIFORNIA

ALTHOUGH SAGEBRUSH is the official flower of Nevada, it was, perhaps, not the best choice, for the rabbitbrush, especially in the fall, is more apt to attract the attention of the casual traveler across the Great Basin. It is so ubiquitous that native Nevadans dependent on the land are likely to have a blind spot for it, except when they focus on its nuisance value. Not uncommonly, however, rabbitbrush along our hard-surfaced roads will be larger and much more vigorous than plants growing further away from the highway. Runoff water is channeled from the road surface to the shoulders, and this phenomenon supplies the rabbitbrush with more water than its neighbors in the adjacent desert receive. A similar effect can be seen in southern Nevada, where conspicuously larger creosote bushes line the highways in such a regular fashion that they appear to have been planted.

Generally, the prevalence of rabbitbrush on rangeland is considered to be an indicator of overgrazing or the result of a decline attributable to some other factor. It is typically a plant of waste areas, abandoned farmsteads, fence rows, and disturbed sites in general. Rabbitbrush resprouts readily after a fire, so that former sagebrush and bitterbrush communities will, within a short time after a fire, appear to be pure rabbitbrush stands. Rabbitbrush is also one of the first shrubs to invade a disturbed area. The small, light seeds possess a tuft of fine hairs which aid in their dispersal by the wind. Good

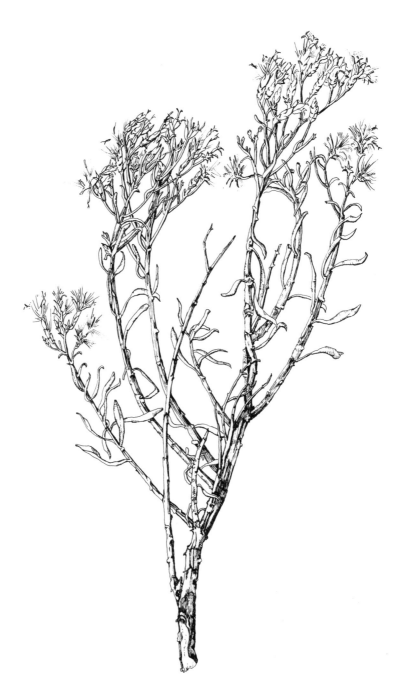

Rubber Rabbitbrush

seed crops are produced virtually every year, and the lack of competition on disturbed sites insures that this shrub will easily become established. Unlike those of many of our native shrubs, the seeds have no afterripening period and do not have to be exposed to cold in order to germinate.

Burgess Kay, Catherine Ross, and Walter Graves studied seed production and germination requirements in the rubber rabbitbrush and found that, while 22 percent of the seeds would germinate immediately after collection from the plant, even after storage at room temperature for a period of thirty months some 10 percent germination occurred. Furthermore, another batch of seeds stored at 4 degrees C. showed no decrease in the germination rate even after forty-three months. Of course, abundant seed production and the ability to grow in disturbed sites are not all the factors involved in the success of rabbitbrush, but, since the seedling stage is extremely critical in the establishment of most plants, it is not surprising that this shrub is eminently successful.

Other names for the rubber rabbitbrush are common rabbitbrush or simply rabbitbrush. "Rubber" is accounted for in a report by S. B. Doten of the Nevada Agricultural Research Service, who related that Great Basin Indians were accustomed to chewing the stems of rabbitbrush in order to extract the latex—the Indians believed that chewing rabbitbrush gum relieved both hunger and thirst. The rubber shortage early in World War II stimulated research on rabbitbrush and other potential rubber-producing plants. Rubber rabbitbrush produces a high-quality rubber, called chrysil, which vulcanizes easily. Nevertheless, it was estimated that the total amount of rubber in all the wild rabbitbrushes in the West would amount to only 30,000 tons, far too little to make the effort of extracting it worthwhile.

Most of the research on rabbitbrush has concentrated on ways of getting rid of it. A variety of methods have been tried, including fire, herbicides, and a procedure called root planing. The latter employs a device pulled through the soil at a depth of 10 centimeters or more, which effectively cuts off the roots below the soil surface. Joseph Robertson and Howard Cords compared various treatments and found that rabbitbrush was a very durable shrub indeed! Sometimes burning would result in a complete kill, but on one occasion 95 percent recovery occurred. Burning combined with the application of herbicides resulted in better control, although the best procedure, albeit the most costly in terms of energy expended, was root planing:

Rubber Rabbitbrush, winter aspect

kills in this case ranged from 80 percent to 21 percent. Survival was apparently poorer when rainfall was low. In any event, the various methods of killing rabbitbrush are so difficult, expensive, and unpredictable that none of us need worry about the species being endangered in any large measure.

The common or rubber rabbitbrush, a moderately branched shrub ranging from 30 centimeters to over 2 meters high, develops from an extensive system of deep roots; typically, it is from 60 centimeters to 1 meter tall. The annually produced, erect, limber branches develop from the stout, woody base, resulting in a broomlike appearance. The leaves—covered with a whitish, feltlike pubescence—vary from 2 to 6 centimeters long. The branchlets are also covered with a similar feltlike pubescence, sometimes so closely compacted that the stems simply appear whitish rather than pubescent.

The bright yellow flowers are borne in rounded clusters at the upper ends of the branches from late summer to fall. Each apparent flower is in reality a cluster of five or six very small, tubular flowers surrounded by a series of bracts. Each flower head is only about a centimeter high. As with many other members of the aster family, the pistil at the base of each flower ripens into a small, dry, hard fruit with a tuft of white or brownish hairs—the pappus—at the upper end. This kind of fruit is known as an achene, and the tuft of hairs represents all that evolution has permitted to remain of the sepals characteristically present in other, more primitive flowering plants.

Rubber rabbitbrush is extremely variable; around twenty subspecies are presently recognized. They differ from one another primarily in the degree of pubescence on various parts of the plant, in the size of the flowering head, and in certain technical details of flower structure. Most of the subspecies have a strong characteristic odor, which is especially apparent when the twigs are broken. They also differ in the habitats occupied: some grow primarily in sagebrush areas, some in alkaline sinks, others in sandy washes, many on rocky slopes, and still others in conifer forests on the intermontane ranges. Probably many of these subspecies have evolved during the last ten thousand years—and, for that matter, many are still evolving. It is readily apparent that some have become physiologically adapted to certain habitats and that some change in appearance has paralleled this adaptation.

Three other species of rabbitbrush are common in the Great Basin, and, while the subspecies are difficult even for the expert, the species are easily distinguished by the amateur. The white-flowered or alkali rabbitbrush, C. *albidus*, is characterized by white flowers and leaves with resinous dots on them. Parry's rabbitbrush, C. *parryi*, resembles the rubber rabbitbrush but has its flowers arranged in a spike at the ends of the branches, and each head has only ten to twenty bracts, compared to twenty to twenty-five for the rubber rabbitbrush. In addition, the bracts in older heads tend to curl downward at their tips. Green rabbitbrush, C. *viscidiflorus*, is easily separated by the lack of a feltlike pubescence on its whitish stems. Usually the branches are smooth, but sometimes a minute pubescence is present, entirely unlike that of the rubber rabbitbrush.

Even though rubber rabbitbrush is of little value to cattle, the flowers and seasonal leaves in late fall and winter are eaten by sheep, goats, and antelope. Deer make some use of this shrub during the winter, though rabbitbrush typically constitutes less than 10 percent of the total food eaten. Birds, rabbits, and other small mammals, as might be expected, make extensive use of the achenes and leaves. Great Basin Indians thought that a tea prepared from the rubber rabbitbrush was good for colds and stomach problems; some individuals, however, show an extreme allergic reaction to rabbitbrush.

The total range of the rubber rabbitbrush, including all the subspecies, extends from eastern California north to British Columbia and east to Saskatchewan, New Mexico, Texas, and the Great Plains. The genus name comes from two Greek words, *chrysos* meaning gold and *thamnos* meaning shrub. There are about fourteen species in the genus, all of them limited to western North America.

Parry's Rabbitbrush
Chrysothamnus parryi

LIKE THE rubber rabbitbrush, which it much resembles, Parry's rabbitbrush has its branchlets covered with a whitish, feltlike pubescence. One major difference is that its flowers are borne in a spikelike affair at the branchlet endings, rather than in a relatively flat-topped or globose cluster, as in the rubber rabbitbrush. Another, frequently conspicuous difference is that the bracts surrounding the flowering head have very long and tapering tips, which tend to curl outward and down, a condition which botanists call squarrose. These same bracts in the rubber rabbitbrush are either blunt or are only moderately tapering, and they are not squarrose.

Parry's rabbitbrush is relatively small, at most only 50 centimeters high. The leaves vary from 1 to 8 centimeters long and from .5 to 8 millimeters wide. As with the green and rubber rabbitbrushes, there are numerous subspecies; about twelve are presently recognized. A number of these are known only from very restricted locales in mountainous terrain. Although the range of habitats and distribution of Parry's rabbitbrush is much like that of the rubber and green rabbitbrushes, the shrub is not as abundant. This latter factor tends to result in smaller populations which are isolated from one another, particularly in mountainous areas. The end result is that chance plays much more of a role in the evolution of new forms, and evolution is, in a sense, speeded up. This is one reason why our mountain ranges and peaks have so many endemic species, species found there and nowhere else.

Another dictum of this scheme of evolution is that natural selection plays a much less important role in such locales than it does in large lowland populations. This implies that many of the morphological differences by means of which we separate these mountain-inhabiting species are not necessarily adaptive. As long as a slight change in form doesn't interfere with the basic physiology of a particular species, it will, in a sense, be tolerated.

One subspecies, *monocephalus*, was originally found by P. B. Kennedy on the summit of Mount Rose, near Reno. However, it is not restricted to that locale; it is known to also occur in Mono and Tuolumne counties in California. This particular form has the flowering spikes reduced to one or two heads. The entire complex of this species ranges from California east to Nebraska and New Mexico.

Although specific studies have not focused on Parry's rabbitbrush, its palatability and value to domesticated animals probably are about the same as those of the green and rubber rabbitbrushes. Unlike the rubber rabbitbrush which it resembles, however, it apparently does not have a significant rubber content. The smaller forms of Parry's rabbitbrush also somewhat resemble species in a related genus known as *Haplopappus* or goldenbush. One species in particular, *H. suffruticosus*, the singlehead goldenbush, superficially might be mistaken for Parry's rabbitbrush but differs in a number of ways. Generally smaller, from 10 to 30 centimeters high, it is not as woody and is more accurately classified as a subshrub, as are most of the other goldenbushes. The branchlets and leaves are covered with a fine glandular pubescence, which is very evident under the hand lens, and the leaves are proportional to their length (1 to 3 centimeters), not as narrow as those of Parry's rabbitbrush; in addition, they have a somewhat crinkled appearance. The flowering heads have three to six strap-shaped or ray flowers, unlike those of any of the rabbitbrushes. Another species, the rubber weed, *H. nanus*, might also be mistaken for some kind of dwarf rabbitbrush, but, again, it has ray flowers and narrow leaves less than 2 centimeters long. It is a small shrub of cliffs and crevices in the arid portions of the Great Basin.

About the only other goldenbush likely to be mistaken for a rabbitbrush is Bloomer's goldenbush, *H. bloomeri*, a compact shrub from 15 to 50 centimeters high with a thick, woody main stem. Its leaves resemble those of Parry's rabbitbrush, and the flower heads are in spikelike arrangements at the ends of the branchlets. Usually the heads have from one to five ray flowers, but sometimes these are lacking. The bracts surrounding the flowering heads appear very different from those of the rabbitbrush; the innermost bracts have a prominent translucent margin, with long hairs all around the edge. The original specimen of Bloomer's goldenbush, from which the description was drawn up by Asa Gray in 1865, was collected on Mount Davidson above Virginia City, Nevada.

The species name *parryi* was assigned to Parry's rabbitbrush by Asa Gray, in honor of Charles Christopher Parry, a nineteenth-century botanist of Iowa and Colorado who served with the Mexican Boundary Survey.

Green Rabbitbrush

Chrysothamnus viscidiflorus

GREEN RABBITBRUSH is easily separated from rubber and Parry's rabbitbrushes by its smooth, hairless, white-barked stems. Typically, it attains a height of less than a meter. Some forms, in fact, are so small that they resemble the snakeweed, but several differences are apparent to the careful observer. For one thing, the small forms of green rabbitbrush usually do not have the broomlike appearance of the snakeweed, and they always lack the strap-shaped or ray flowers present in each flowering head of the snakeweed. Additionally, the fruiting bodies in the snakeweed are topped by tiny, pointed scales visible under a hand lens, rather than the tuft of hairs characteristic of the rabbitbrushes. On the other hand, some of the larger forms of green rabbitbrush are so different in appearance as to be scarcely recognizable as the same species.

Green rabbitbrush has been described as an extremely polymorphic species, one with many forms. There are five subspecies recognized today, though at one time many more subspecies, varieties, and forms were considered to exist. Most of these are not regarded as valid today because of the numerous intergrades which exist between them. Undoubtedly, we are presented here with the same problem we have seen in other Great Basin shrubs: a rapidly evolving complex of forms which developed as a result of the new habitats that opened up following the last ice age. Perhaps a few thousand years from now some of these forms will be distinct species, but that is not likely to occur as long as they interbreed freely.

The various subspecies are separated from one another on the basis of height, leaf size and shape, and the occurrence at times of a minute amount of pubescence on the leaves. The leaves, typically green and shiny or viscid, are sometimes dotted on the upper surface with tiny, clear, glandular spots.

Sometimes the larger forms have the leaves twisted lengthwise, so much so that at one time a variety *tortifolia* was described. Some leaves are narrow, lance-shaped affairs 6 centimeters long, while the narrow leaves on the smaller varieties are only a millimeter wide and 1 or 2 centimeters long.

James Young and Raymond Evans, of the Agricultural Research Service at Reno, studied the development of the green rabbitbrush in detail in order to discover the best time to apply herbicide. The plants studied were at a site about 35 kilometers north of Reno which had been burnt some thirteen years previously. The only woody shrub which had become reestablished in that time was the green rabbitbrush. Young and Evans found that growth in the spring and summer could be divided into four stages: a period when the buds noticeably swelled, called bud burst; a period of slow, restricted growth; a period of rapid, accelerated growth; and the cessation of growth, when flower bud development took place. Throughout March and April only very restricted growth occurred, probably because of the low soil temperatures. During May there began a period of accelerated growth which lasted until July. During a very dry year, Young and Evans found that growth was consistently very slow throughout the season, with no period of acceleration. Shrubs were divided into four age classes from seedling to senescent, and, as might be expected, seedlings and young plants experienced a more rapid and a greater amount of growth than mature or senescent plants. Flowering was first apparent in mid August, with the plants being in full bloom by the first of September. Within a week or two, the achenes were fully formed and had begun to fall from the plant, although some persisted until spring. Many achenes, however, aborted for undetermined reasons. Nevertheless, Young and Evans calculated that the average rabbitbrush produced about thirty thousand achenes, and at the site studied this meant about 20 million achenes per hectare per year!

In any event, the observations by Young and Evans showed that the best time to apply 2,4-D as a herbicide was during the period of accelerated growth. Application before or after that period had a much less damaging effect on green rabbitbrush. They cite an example of a big sagebrush area sprayed with 2,4-D, which killed the big sage but failed to control the rabbitbrush. As a result of the release from competition, rabbitbrush rapidly took over the area—it required some thirteen years for sagebrush to begin to reinvade the stand. One natural control on rabbitbrush populations ap-

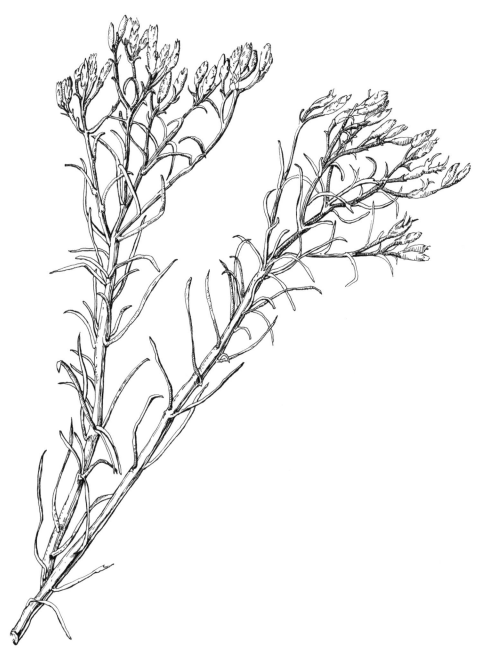

Green Rabbitbrush

peared to be the larvae of an insect of the genus *Aemaeodera*, which burrowed into the stems. Young and Evans found that almost every large stem sectioned showed tunnels produced by these larvae. They concluded that a steady decline in the population at this site was occurring as a result and that this prepared the way for the invasion of big sagebrush, the next stage in succession.

In another study at the same site, Young and Evans found that, in areas where big sagebrush was destroyed by fire or herbicide, root sprouting of green rabbitbrush occurred; these plants then served as the source of an enormous number of achenes, so that the burnt or sprayed areas rapidly developed dense stands of rabbitbrush. This particular site, known as Medell Flat, was characterized by two investigators, P. B. Kennedy and S. B. Doten, as a virtual dust bowl in 1901, due to overgrazing by sheep—John Muir's "locusts" of the Sierra Nevada. Grazing controls implemented in 1935 allowed some shrub vegetation to return, albeit not to the conditions that must have existed before grazing began in the middle of the nineteenth century. Almost certainly, many more perennial grasses, such as the Great Basin wild rye, were abundant before the settlers came. For example, Paul Tueller and Raymond Evans found that green rabbitbrush could be effectively controlled by a herbicide called picloram and that, after sagebrush and rabbitbrush were killed, the crested wheatgrass population increased significantly.

Although quantitative estimates are hard to come by, the landscape of the Great Basin, except for the alkaline flats and shadscale deserts, was formerly much grassier than it is now, as natural as sagebrush lands may seem to casual observers today. The constant pressure of more than a century of grazing has favored the development and probably the evolution of shrubs, with the result that humans have been forced to create artificial ways of controlling this excess population. In short, just as the retreat of the Ice Age opened up new niches for many plants, humans have produced new niches for shrubs by reducing the competition they formerly had to contend with from grasses.

In the eastern part of the Basin, a related species of rabbitbrush occurs which has white flowers. This form—the white-flowered or alkali rabbitbrush, *C. albidus*—extends westward across to Death Valley but is rare in that portion of its range. In addition to its flower color, it differs from the green rabbitbrush, which it otherwise resembles, in that the small bracts which surround each flowering head have a slender, tapering point rather than the blunt or acute tip of the green rabbitbrush.

The palatability of green rabbitbrush, as far as domestic animals are concerned, is about the same as that of the rubber rabbitbrush, and it is utilized to a similar limited extent. Indians made use of the plant as a treatment for colds and as a poultice for rheumatism. Percy Train reported that finely mashed leaves were inserted into tooth cavities to stop toothaches, though with what success he doesn't say!

Green rabbitbrush is characteristic of sagebrush and pinyon-juniper associations. Like the rubber rabbitbrush, it can also be found in alkaline areas and at high elevations in the Great Basin. Its total range is from southern California to British Columbia and east to New Mexico, Montana, and the Rocky Mountains.

Other names for the green rabbitbrush are rabbitsage, yellowbrush, and yellowsage. The botanical name means sticky- or viscid-flowered, though the flowers are not noticeably so. It is really more sticky-leaved than sticky-flowered—the scientific name is somewhat of a misnomer.

Snakeweed
Gutierrezia sarothrae

SNAKEWEED or broom snakeweed is most commonly mistaken for a small rabbitbrush, especially when in bloom. However, it is easily distinguished by its numerous erect, slender stems, much thinner than those of rabbitbrush, and by its flowering heads, which have strap-shaped or ray flowers around the edge as well as tubular disk flowers in the center. Rabbitbrush has no ray flowers and thicker as well as fewer stems. The seedlike structures or achenes which develop from the pistils of snakeweed have four to fifteen oblong, ragged scales at their upper ends; these can easily be seen with a hand lens. In the rabbitbrush, the achenes have numerous fine, soft, white or brownish hairs at the upper ends.

Other names for snakeweed which are quite descriptive of its personality are matchweed, turpentine-weed, and yellow top. The ecological range of snakeweed in the Great Basin is wide and varied. It can be found abundantly on slopes with big and dwarf sagebrush as well as down in the shadscale desert association. On the Great Plains it can be found with grama and buffalo grasses, and in the southwest deserts its companions may be creosote bush and mesquite. Soil habitats range from shallow, sandy or rocky situations to clay loams, although snakeweed seems unable to grow in saline or alkaline environments.

Snakeweed gets to be 20 to 60 centimeters tall. The leaves are about 2 millimeters wide, 10 to 35 millimeters long, shiny green, and somewhat sticky to the touch. Snakeweed has a deep taproot, as well as many long lateral roots which are able to make use of rains which moisten only the top layers of soil. The small flowering heads, about 3 millimeters across, contain several disk and several ray flowers—usually no more than eight altogether, however.

Snakeweed

Snakeweed is distributed from southern California north to eastern Washington, Manitoba, and Montana and south through the Great Basin, Rocky Mountains, and Great Plains to Arizona, New Mexico, and Texas. Ranges which have been overgrazed are aggressively invaded by snakeweed, since in many areas it is considered worthless as cattle fodder. However, parts of Nevada and Utah are so depauperate as cattle ranges that in such locales snakeweed is regarded as fair for sheep, at least during the winter, and poor for horses and cattle. Apparently, heavy usage of snakeweed by cattle may sometimes result in poisoning and death. It is common in parts of the Great Basin to see snakeweed appearing in abundance a few years after a severe sagebrush burn.

The genus contains about twenty species, all in the New World and most native to western North America. The genus name honors a Spanish family, Gutiérrez, while the species name is from the Latin *sarothrum* and means broom. Some botanists now place snakeweed in the genus *Xanthocephalum*, though arguments still persist as to which name is correct.

White Burrobush
Hymenoclea salsola

ALONG SANDY WASHES in the southern Great Basin there commonly grows a tall shrub which, on occasion, seems to be decorated with pearls. To some it is known as the desert pearl or pearlbush. These pearls are formed by pea-size clusters of papery, translucent bracts fused at the base into a woody affair which surrounds the female flowers. The narrow, often threadlike leaves on light tan stems are another distinguishing feature. On closer examination, some of the leaves produced during the spring may be seen to consist of several or more divisions. Under a hand lens, the narrow leaves can be seen to possess a very unusual feature: a lengthwise groove on the upper surface covered with hairs. The remainder of the leaf surface often lacks any pubescence.

Like big sagebrush, white burrobush belongs to the aster family, but it differs significantly from big sagebrush in that the stamens and pistils are borne in separate flowers and these, in turn, are in separate clusters. It shares this particular feature with the ragweeds (*Ambrosia*) and cockleburs (*Xanthium*), to which it is closely related. Like the ragweeds it is wind-pollinated and, unfortunately, it is also a significant hay fever plant. In the white burrobush, the bracts surrounding the female flowers are relatively thin and winglike, fused at the base, and make up the characteristic pearl. In the ragweeds and cockleburs, on the other hand, the bracts, while similarly fused at the base, develop instead into spines which are frequently hooked. Kathleen Peterson and Willard Payne, who carried out a comprehensive study of this genus, concluded that it evolved from an ancestor common to the ragweeds and cockleburs, that the primitive pistillate cluster of bracts probably was somewhat like that of some *Ambrosias*, and that the present arrangement of the bracts is the result of an evolutionary process

White Burrobush

called neoteny. Wings and spines of the pistillate clusters of all *Hymenoclea, Ambrosia,* and *Xanthium* species are upright and appressed during the early developmental stages. The burrobushes appear to have retained this juvenile feature into adulthood. Additionally, Payne proposes that this whole group of ragweeds, cockleburs, and burrobushes first evolved somewhere in the arid Southwest and from there spread elsewhere.

Neoteny is said to occur when a juvenile or embryonic structure is retained into the adult state. This developmental process is poorly understood, but apparently it involves some mechanism which stops differentiation before normal maturity has been reached. One example of this among animals is the Mexican axolotl, a type of salamander which, under certain conditions, matures and breeds while still in the larval form, with the characteristic external gills of that stage. It is thought by some experts on the evolution of higher plants that neoteny played a very important role in the evolution of flowers from the conelike reproductive structures of their primitive precursors over 70 million years ago. Some think that neoteny serves to simplify structures and, in a real sense, free up genes which are then available for other possible roles in evolution or, to be more precise, in individual development as determined by evolution.

To get back to the white burrobush, the bracts on the pistillate flower clusters serve to disseminate the seeds by wind or, undoubtedly, by water as well, since this is such a common shrub along washes. In the case of the cockleburs, the hooked fruits disseminate by becoming entangled in the fur of animals and, not infrequently, in human clothing. The clusters of staminate flowers are borne on the same branches as the pistillate flowers and adjacent to them; as the seeds develop, the staminate flowers drop from the plant. The aboveground portions of the white burrobush are relatively short-lived. Flowers are borne on two-year-old branches which, following fruit development, die back to the ground. The roots undoubtedly live many more years, though no one has attempted to find out how many.

Frank Vasek, H. B. Johnson, and D. H. Eslinger studied the effects of pipeline construction on creosote scrub vegetation in the Mohave Desert of southern California. They found that the most abundant pioneer shrub was the white burrobush, which in some disturbed areas made up as much as 85 percent of the vegetation cover twelve years after the original vegetation had been removed. However, they estimated that, because of the slow growth of the creosote bush and its long-lived associates, the reestablishment of the original vegetative cover might well take centuries or millennia. Of all the vegetation types, desert plants are certainly one of the most fragile and most easily capable of being permanently altered or destroyed by human activity.

The genus name *Hymenoclea* means closed membrane in the original Greek; it refers to the fused bracts surrounding the flowers. The species name *salsola*, derived from the Latin, means salty. Prior to the work of Peterson and Payne, *Hymenoclea* was thought to comprise four species and two varieties. They found, however, that there were only two species in the genus. One of these, *monogyra* or the singlewhorl burrobush, differs in its more treelike shape as well as its fall blooming period, rather than spring, as is the case with the white burrobush. The singlewhorl burrobush is common in the Chihuahuan and Sonoran deserts from Texas to California and south into Mexico.

Three varieties are now recognized as making up the species *salsola*, based on the nature of the wings forming the pearls. One of these, known as variety *patula*, has a northern distribution well into the Great Basin; another, variety *pentalepis*, occurs further to the south; and the overlap in the range between these two varieties is occupied by the intermediate form known as variety *salsola*. Just as we saw in big sagebrush, it appears that the differing physiologies of the various races so far recognized are reflected in slightly different appearances. In effect, we are once again watching evolution at work.

Gray Horsebrush
Tetradymia canescens

LIKE THE littleleaf horsebrush, the gray horsebrush occurs commonly throughout the Great Basin. Generally, although it is a plant of dry habitats, it is found in relatively less dry situations than the littleleaf. It is common in sagebrush, pinyon-juniper, and even ponderosa communities up to an elevation of 8,000 feet. Also, unlike the littleleaf, occasional dense stands of gray horsebrush occur within the sagebrush zone. Its distribution ranges from southern California, Arizona, and New Mexico north through Nevada and Utah to British Columbia, Montana, Wyoming, and Colorado.

Gray horsebrush is a plentifully branched shrub that attains a height of between 30 and 60 centimeters. It is easily separated from the littleleaf horsebrush by its grayish-pubescent young stems and leaves; the latter are soft, narrow, and between 12 and 25 millimeters long. Gray horsebrush does not tend to produce clusters of leaves at the leaf axils, as is the case with the littleleaf horsebrush. The beginner may at first mistake it for some kind of rabbitbrush, but the flower heads, aside from being produced earlier in the year, have the characteristic four or five woody bracts surrounding them, unlike the many smaller bracts typical of the rabbitbrushes.

The flowering heads, produced from June into September, are borne in small, flat-topped clusters very much like those of the littleleaf horsebrush. Oddly enough, flowering begins first in British Columbia, and the last shrubs to flower inhabit southern California and Arizona. Individual flowers are similar to those of the littleleaf, described under that species. The fruiting structures or achenes are not always heavily pubescent, as are those of the littleleaf; on occasion, they may be completely smooth.

Both the gray and the littleleaf horsebrush are of virtually no value as forage for domestic animals. Indeed, sheep have been known to develop bighead disease as a result of eating too much of either shrub—the accumulation of fluid in the head in such cases can lead to death. No one has yet identified the poisonous principle involved, but it is known that consumption of leaves and stems amounting to as little as one-half of one percent of the animal's body weight is enough to cause liver damage and an associated sensitizing to light. This light sensitivity may then result in capillary damage and swelling of the head. Sheep kept in the shade after eating horsebrush are less susceptible to the disease, and one study has shown that in some cases the light sensitivity caused by horsebrush occurs only after sheep have also consumed *Artemisia arbuscula* variety *nova*, black sagebrush. Cattle appear not to be similarly affected by horsebrush, according to A. B. Clawson and W. T. Huffman, who studied this disease in the 1930s. Indians are variously reported to have used extracts from gray horsebrush as a physic and as a treatment for venereal disease.

The derivation of the genus name, *Tetradymia*, is covered in our discussion of the littleleaf horsebrush. There are ten species in the genus, all of them confined to the Southwest. *Canescens* comes from the Latin *canesco*, which means to become white or hoary—plants which have a fine, dense, gray or white pubescence are said to be canescent. John Strother of the University of California at Berkeley, who has studied this genus intensively—examining over twenty-eight hundred specimens in the laboratory and in the field—considers that the gray horsebrush is a kind of ancestral prototype of the genus. It appears to be the least adapted to a desert situation. The evolution of anatomical features that allowed the horsebrushes to survive in drier habitats has occurred as the Southwest became drier over recent geological time.

The common name horsebrush has an obscure origin. The shrub is certainly of no value as forage for horses, though deer are said to eat the young shoots and leaves. Another name, infrequently used, is black sage. Horsebrush, of course, aside from belonging to the aster family, as does sagebrush, is not at all closely related to sagebrush.

Littleleaf Horsebrush
Tetradymia glabrata

ONE OF OUR MOST attractive desert shrubs when in flower is the littleleaf horsebrush—its flat-topped clusters of yellow flowers are particularly conspicuous, even when seen from a distance, from April through July. If we ever really get around to making use of our desert shrubs for landscaping, in order to conserve water, the littleleaf horsebrush would be an ideal candidate. It is an early riser, frequently beginning to leaf out in February when most other shrubs are still dormant. By midsummer, the leaves on bushes in drier areas will have turned brown. Other shrubs may be only beginning to develop flower buds, but littleleaf will be finished for the year. It is an opportunist, however, and if a wet summer occurs it will remain green and even produce new leaves and shoots if there is enough water.

Common throughout the Great Basin as a scattered shrub in the shadscale desert and sparingly on up into the sagebrush zone, littleleaf extends outside of this area to central Oregon, central Idaho and Montana, and south into the Mohave Desert. Other communities in which it is found include rabbitbrush and greasewood. It appears from this to be considerably more ecologically tolerant than many of our other desert shrubs.

Littleleaf horsebrush is a moderately branched shrub between 30 centimeters and 1 meter tall. Most of the branches are ascending and are characteristically white-pubescent, except for the narrow, short, smooth streaks below each leaf node. This is a feature which the littleleaf shares with the gray and Nuttall's horsebrushes (described in our discussion of the shortspine horsebrush). Aside from these three species, no other Great Basin shrub has this particular characteristic of alternating smooth and pubescent areas on its stems. The first leaves to appear at each node are rigid, awl-shaped affairs between 8 and 12 millimeters long. These are lost fairly early and are suc-

Littleleaf Horsebrush

ceeded by clusters of soft, narrow, elongate, and succulent leaves about the same length or a little longer. These clusters are located on the upper side of the angle of each of the first leaves, a location termed the axil. The axillary leaves are a prominent bright green and smooth at maturity.

The beginning student of our local flora may not at first recognize littleleaf horsebrush as a member of the aster family, for unlike most members of that family, which have many flowers congested into a head, there are only four flowers in a single head. And, later in the season after blooming has occurred, the four hard bracts surrounding these flower clusters resemble nothing so much as a capsule split four ways to release the four hairy seeds. Each of these is in reality not a seed but a structure called an achene, which consists of a ripened and woody pistil that encloses a single seed; the unhulled sunflower "seed" is a good example of an achene. Examination of the yellow flowers under a hand lens will show that each consists of a tubular corolla that is five-lobed at the top, the whole being perched on top of the pistil. One can also see very long, fine, upward-pointing, straight hairs—the pappus—situated at the juncture of the corolla and the pistil. This pappus is all that is left of the sepals, we believe.

Because of their low palatability and scattered distribution, horsebrushes are regarded as poor to useless where cattle and horses are concerned, so much so that the comprehensive *Range Plant Handbook* of the U.S. Forest Service, published in 1937, fails even to note the shrubs' existence.

The genus name *Tetradymia* comes from two Greek words, *tetra* which means four and *dymos* which implies together. Strictly speaking, the concept doesn't apply to all horsebrushes, since some have up to nine flowers in a head. *Glabrata* is a Latin term for smooth or hairless; in this case, the reference is to the glabrous leaves. From an evolutionary standpoint, the horsebrushes show something we have already discussed earlier, that is, the inclination to reverse a previous trend. We consider the aster family to be very advanced, since for many of its members reproduction involves the fusion of the supporting structures for a large number of flowers in the form of a head surrounded, frequently, by green leafy bracts. A good example of this very complicated arrangement would be the sunflower head. In the horsebrushes, this trend in evolution appears to have been reversed, with the result that only a few flowers remain in each head. In fact, in some members of the aster family, this reduction in the number of flowers reaches the ex-

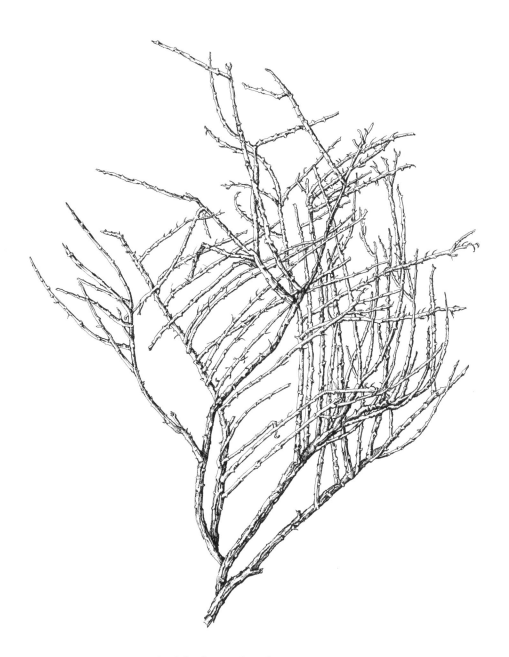

Littleleaf Horsebrush, winter aspect

treme of only one flower per head, as is characteristic of some of the bur sages in the deserts to the south of our area.

An even more remarkable evolutionary pattern occurs in many members of the wire-lettuce genus, *Stephanomeria*, found throughout our desert areas. These species have the number of flowers reduced to five per head. The characteristic tubular flower in these forms, in common with all of the lettucelike members of the aster family, is split to the base and flattened out into a ribbonlike structure. These five ribbons project radially so that, to the uninitiated, there appears to be one single flower with five separate petals. The illusion is completed by the five bracts below each head, which appear to be five sepals!

For those who like to believe that evolution works consistently toward a goal, this situation must present an enigma. It is easier to make sense of situations like this if one assumes that the evolution of new forms is simply opportunistic—sometimes things become more complex, sometimes they become simpler, but inherently there is a random aspect to it all. Why is five such a magic number for flowering plants? Probably because bees or other pollinators are pretty good at geometry and can distinguish a radial pattern of five and separate it easily from three, four, or six. In other words, this is another signal to the potential pollinator, aside from the color and general symmetry of the flower. Obviously this can't be the entire story, for the pattern also has something to do with the complex physiological factors, still only poorly understood, that determine how many leaves occur at a node, what the spiral arrangement will be, and so on.

Shortspine Horsebrush
Tetradymia spinosa

SHORTSPINE OR catclaw horsebrush is a rigid, abundantly branched shrub ranging from half a meter to a meter tall. Its branches are prominently covered with white, woolly, densely matted hairs, accounting for another common name, cottonthorn. As is the case with littleleaf horsebrush, shortspine is common all over the Great Basin but never occurs in pure stands—usually it is found as isolated individuals or small colonies scattered here and there through the shadscale desert. John Strother, in his studies on the horsebrushes, notes that the section of the genus which includes the shortspine and cotton horsebrushes is characterized by shrubs which produce rhizomes running out to 30 centimeters from the parent plant. These allow the plant to reproduce vegetatively, resulting in colonies with a great deal of genetic uniformity. Typically, the individuals in a colony bloom at the same time, have similar leaf and stem characteristics, and so on. Although frequently found with littleleaf, in general shortspine inhabits the drier flats and slopes in desert areas.

We noted that the first littleleaf leaves to develop were rigid, awllike structures. In the shortspine this evolutionary development has been carried one step further: these first or primary leaves have been modified into short, straight or usually curved spines 2 centimeters or less in length. And, like littleleaf, the secondary leaves are borne in clusters at the leaf axils. These secondary leaves are hairless—or nearly so—narrow, cylindrical affairs a centimeter long or less. The flowering heads occur in small, flat-topped clusters at the ends of the younger branches and closely resemble those of the littleleaf. However, whereas the latter typically has four flowers in each head, the shortspine has six or seven, and has five or six hard, woody bracts enclosing the head, rather than only four as in littleleaf. The achenes resemble those of littleleaf but are much more densely hairy.

It is pretty clear that littleleaf and shortspine horsebrushes had a closely related common ancestor. Shortspine shows some of the adaptations we have come to expect in desert plants: an increased pubescence, the development of spines, and a tendency toward succulent leaves. There is no evidence that it is of any value to domesticated animals, and apparently it was little used by native Americans, though there is one report that the spines were once used as tattoo needles!

Another closely related species, the longspine horsebrush, *T. axillaris*, occurs in southern Nevada, southern Utah, Arizona, California, and the western Great Basin north to southeastern Oregon. It differs from shortspine primarily in that its spines are usually straight and from 2 to 4 centimeters long. Some authorities have considered this species as only a variety of the shortspine. Strother's studies, however, show that the longspine is definitely a distinct species. The shortspine's total range includes part of California immediately adjacent to central Nevada, southeastern Oregon east to southwestern Montana, and Utah and western Colorado.

Still another spiny horsebrush, which occurs in the eastern Great Basin in Utah and adjacent Nevada, is Nuttall's horsebrush, *T. nuttallii*. It can be recognized by the alternate smooth and pubescent areas on the stems, like those of the littleleaf and gray horsebrushes. In addition, the achenes have seventy-five to one hundred fine, white bristles at their upper ends, compared to only twenty-five flattened, white, narrow structures in the shortspine horsebrush. More obvious, however, is the pubescence on the shortspine achenes, which consists of fine, white hairs up to 1 centimeter long. The achenes in Nuttall's horsebrush have only a short pubescence. The latter has spines similar to those of the shortspine—.5 to 2.5 centimeters long. The easiest way to separate the two, however, is to examine the stem for the smooth zones.

Cotton Horsebrush
Tetradymia tetrameres

COMMON ON stabilized sand dunes or deep sand in the western Great Basin, the cotton horsebrush is our tallest example of the genus, reaching heights up to 2 meters. Characteristically, it has few branchlets on its long, wandlike, white stems. The very heavy, compact pubescence which covers the stems accounts for its common name. The first seasonal leaves are very narrow and up to 3 centimeters long, while the secondary leaves are somewhat shorter, between 1 and 2 centimeters in length. The short-lived leaves soon disappear with the hottest part of the summer, so that the shrub's usual aspect is one of leafless, long, white branches projecting from the top of a dune. Cotton horsebrush is also able to reproduce vegetatively by means of long, horizontal rhizomes extending out from the main plant.

The flowering heads, produced from June into July, are borne in clusters of four to six on short side branches. As in littleleaf horsebrush, there are four woody bracts surrounding each flowering head, which in turn consists of just four individual flowers. The hard achene produced by each flower is covered with a fine, white pubescence and has a ring of stiff, white bristles at the upper end.

For a long time the cotton horsebrush in the western Great Basin was assigned to the species *comosa*. However, John Strother has shown that *comosa* is strictly a southern California plant, separated from our cotton horsebrush by nearly 400 kilometers. *Comosa* differs in several ways—it has five to eight flowers in each head, wider leaves, and achenes lacking the stiff, white bristles. Once experienced, the not unpleasant odor of a freshly broken cotton horsebrush branch is easily recognized and can be used to identify this shrub.

One interesting facet of the *Tetradymia* flowering pattern is that the individuals in a given colony all produce flowers within a few days of one another. A variety of insects visit the flowers—beetles, flies, bees, and moths. Pollination is also encouraged by the fact that the horsebrushes are usually the first desert shrubs to flower. Of our species, the littleleaf horsebrush is the first to bloom, while flowering in the gray horsebrush may continue into September in the southern part of its range outside the Great Basin. The latter species, not surprisingly, is the least adapted to desert conditions. Another distinctive aspect of the gray horsebrush is that flowering begins in British Columbia in mid June and progresses southward. In the other species the pattern is reversed: in littleleaf horsebrush, for example, flowering begins in April in the Mohave Desert of California and doesn't begin until mid June in Idaho.

It might be assumed that, because the cotton horsebrush grows on dunes, it is adapted to a very dry habitat. However, like those of many dune plants, its roots extend far down into moister zones within the sand. Dunes, in general, are good accumulators of rainfall, since their porous aspect allows water to penetrate rapidly without any runoff. In addition, because the pores are so large, capillary action that might otherwise bring the water to the surface to evaporate during dry periods is insignificant.

The species name *tetrameres* means having four members in a whorl—obviously a reference to the structure of the flowering heads. Outside of Nevada, the Great Basin cotton horsebrush extends only to Mono County in adjacent California.

Indian Names for Great Basin Shrubs

NOTE the following abbreviations: NP/Northern Paiute; SP/Southern Paiute; SH/Shoshone; W/Washoe.

Acer glabrum   SP: pagwiabɨ "water oak"
Allenrolfea occidentalis   NP: kuhána
Alnus tenuifolia   SH: uguzupɨ
Amelanchier alnifolia   NP: tɨabui; SP: tɨwampi; W: šuʔwetik; SH: tɨʔampi
Arctostaphylos patula   NP: tɨmaiya; W: ʔeyeyeʔ; SH: tɨmaiyɨha "to mix," eg. with tobacco
Artemisia arbuscula   NP: tɨbisiginupɨ
A. spinescens   NP: kɨɨbatɨkanogwa "ground squirrel food"
A. tridentata   NP: sawabi; SP: saŋwabi; SH: pohobi; W: daabal
Atriplex canescens   NP: taʔɨbi
A. confertifolia   NP: kaŋubɨ
Berberis repens   SP: tɨwikunukwi "stomach medicine"
Ceanothus velutinus   SH: adadɨmpibɨ
Ceratoides lanata   NP: sɨssubɨ
Cercocarpus intricatus   NP: tuupi; W: duhul; SH: tunambi
Chrysothamnus nauseosus   NP: sigupi; SP: sɨkɨmpɨ; SH: sibupi
C. viscidiflorus   NP: isisigupi; W: boop'oʔ
Cornus sericea   SH: eŋqakwinubɨ
Cowania mexicana   SP: ɨnapɨ; SH: piahɨnabi "big bitterbrush"
Ephedra nevadensis   NP: cudupi; SP: tutupi; W: meegel; SH: tudumbi
E. viridis   NP: cudupi; SH: tudumbi
Grayia spinosa   W: balŋat'saŋ
Holodiscus dumosus   SH: toyahɨnabi "mountain bitterbrush"
Juniperus communis   NP: waapi; SP: waʔapi; W: p'aal; SH: waapi
Kochia americana   NP: wazobɨ
Lepidium fremontii   SP: aka
Lonicera involucrata   SH: wɨ-

319

dandɨkapɨ "bear's food"
Lycium shockleyi　NP: huupi; SP: uʔupi
Prunus andersonii　NP: canabi; W: c'ipapa
P. emarginata　NP: sɨɨbi "willow;" was used like willow as frame for baby basket
Purshia tridentata　NP: hɨnabi; W: balŋat'saŋ; SH: hɨnabi
Rhus diversiloba　W: dap'ap
R. trilobata　SP: iʔiši or sɨʔɨbi
Ribes aureum　NP: wogobissa; W: nanholwa; SH: ohapogombi
R. cereum　SH: engapogombi
R. velutinum　NP: muguciabi; W: maʔkiʔʔiyek; SH: mugubogombi
Rosa woodsii　NP: ciabui; SP: ciampibɨ; W: pet'sumeliʔ; SH: ciʔabi
Salix exigua　NP: sɨɨbi; SP: kanabɨ; W: himu; SH: sɨhɨbi
Salvia dorrii　W: p'oʔlo p'isew "rat's ear"; NP: kanigiaʔa
Sambucus cerulea　NP: hubu; W: baaduʔ
S. melanocarpa　NP: hubu; SP: kunukwi; SH: tɨiyambɨʔɨ
S. racemosa　NP: isabui; SH: konogipɨ
Sarcobatus baileyi　NP: tonobi; SH: tonoʔobi
S. vermiculatus　NP: tonobi; SH: tonobɨ
Shepherdia argentea　NP: wiyɨpui; SP: paʔoipi; W: dawal; SH: wiʔyɨmbi
Suaeda sp.　NP: wazobɨ; SP: adɨmpɨ; SH: atɨmpɨ
Symphoricarpos sp.　SP: tampisudupi
Tamarix sp.　SP: pawaapi "water juniper"
Tetradymia canescens　NP: sigupi; SH: sigupi
T. glabrata　NP: wacidinubɨ
T. spinosa　NP: togogwa tama "rattlesnake's teeth"
T. tetrameres　SH: kusisigupi "gray rabbit's brush"
Vaccinium uliginosum occidentale　NP: tokabonoma

Vowels:
ɨ between i and u not found in English.
a as a in f<u>a</u>ther.
e as in b<u>ai</u>t.
u as in b<u>oo</u>t.
o as in t<u>oe</u>.
i as in s<u>ee</u>.

Consonants:
values the same as in English except for:
c as in ca<u>t</u>s.
ʔ glottal catch as in oh-oh.
ŋ ng as in si<u>ng</u>.
t', p', etc. are glottalized consonants without English equivalents. Long consonants, such as ss, are held longer. The same is true of long vowels, such as aa.

BIBLIOGRAPHY

Abrams, Leroy. 1940–1960. *Illustrated Flora of the Pacific States.* Vols. 1–4. Stanford Univ. Press, Stanford.

Anonymous. 1937. *Range Plant Handbook.* Forest Service, U.S. Dept. of Agr., Washington, D.C.

———. 1957. *Nevada, a Guide to the Silver State.* American Guide Series. Binfords and Mort, Portland.

Bailey, Liberty H. 1900–1902. *Cyclopedia of American Horticulture.* 4 vols. Macmillan, New York.

Baker, G. A., P. W. Rundel, and D. J. Parsons. 1982. Comparative phenology and growth in three chaparral shrubs. Bot. Gaz. 143:94–100.

Bakker, Elna. 1971. *An Island Called California.* Univ. of Calif. Press, Berkeley and Los Angeles.

Baum, Bernard R. 1967. Introduced and naturalized tamarisks in the United States and Canada (Tamaricaceae). Baileya 15:19–25.

Beetle, Alan A. 1960. A study of sagebrush: the section *tridentatae* of *Artemisia.* Univ. of Wyoming Agr. Exp. Sta. Bull. 368.

———. 1971. An ecological contribution to the taxonomy of *Artemisia.* Madrono. 20:385–386.

Benson, Lyman, and Robert A. Darrow. 1981. *The Trees and Shrubs of the Southwestern Deserts*. Univ. of Arizona Press, Tucson.

Bessey, Charles E. 1915. The phylogenetic taxonomy of flowering plants. Ann. Mo. Bot. Gard. 2:109–164.

Bidwell, G. L. 1925. Saltbushes and their allies in the United States. U.S. Dept. of Agr. Bull. 1345.

Billings, W. Dwight. 1945. The plant associations of the Carson Desert region, western Nevada. Butler Univ. Bot. Stud. 7:89–123.

———. 1951. Vegetation zonation in the Great Basin of western North America. In *Les bases ecologiques de la régénération de la végétation des zones arides*, Intl. Colloquium, Intl. Union of Biol. Sci., Ser. B, 9:101–122.

Bissell, H., B. Harris, H. Strong, and F. James. 1955. The digestibility of certain natural and artificial foods eaten by deer in California. Calif. Fish and Game 41:57–58.

Blackwell, Will H., Jr., Mark D. Baechle, and Gene Williamson. 1978. Synopsis of *Kochia* (Chenopodiaceae) in North America. Sida 7:248–254.

Brown, Grant D. 1956. Taxonomy of American *Atriplex*. Amer. Midl. Nat. 55:199–210.

Caldwell, M. M., and L. B. Camp. 1974. Belowground productivity of two cool desert communities. Oecologia 17:123–130.

Carpenter, Philip L., and David L. Hensley. 1979. Utilizing N-fixing woody plants species for distressed soils and the effect of lime on survival. Bot. Gaz. 140 (suppl.):S76–S81.

Christensen, Earl M. 1962. The rate of naturalization of *Tamarix* in Utah. Amer. Midl. Nat. 68:51–57.

Cronquist, Arthur, Arthur H. Holmgren, Noel H. Holmgren, and James L. Reveal. 1972. *Intermountain Flora*. Vol. 1. New York Botanical Garden, New York.

———. 1981. *An Integrated System of Classification of Flowering Plants*. Columbia Univ. Press, New York.

Davis, Craig B. 1973. "Bark striping" in *Arctostaphylos* (Ericaceae). Madrono. 22:145–149.

Dawson, Jeffrey O., and John C. Gordon. 1979. Nitrogen fixation in relation to photosynthesis in *Alnus glutinosa*. Bot. Gaz. 140 (suppl.): S70–S75.

DeBell, D. S., and M. A. Radwan. 1979. Growth and nitrogen relations of coppiced black cottonwood and red alder in pure and mixed plantings. Bot. Gaz. 140 (suppl.): S97–S101.

DePuit, E. J., and M. M. Caldwell. 1973. Seasonal pattern of net photosynthesis of *Artemisia tridentata*. Amer. J. Bot. 60:426–435.

DeQuille, Dan. 1963. *Washoe Rambles*. Westernlore Press, Los Angeles.

Diettert, R. A. 1938. The morphology of *Artemisia tridentata* Nutt. Lloydia 1:3–74.

Dina, Stephen J. 1970. An Evaluation of Physiological Response to Water Stress as a Factor Influencing the Distribution of Six Woody Species in Red Butte Canyon, Utah. Thesis (Ph.D.), Univ. of Utah, Salt Lake City.

Doten, S. B. 1942. Rubber from the rabbitbrush (*Chrysothamnus nauseosus*). Nev. Agr. Exp. Sta. Bull. 157.

Earle, Alice Morse. 1971. *Sun-Dials and Roses of Yesterday*. Orig. ed., 1902. Tuttle, Rutland, Vt.

Eckert, Richard E. 1954. A study of competition between whitesage and *Halogeton* in Nevada. J. Range Mgmt. 7:223–225.

——— and Raymond A. Evans. 1968. Chemical control of low sagebrush and associated green rabbitbrush. J. Range Mgmt. 21:325–328.

Ehleringer, J. R., and H. A. Mooney. 1978. Leaf hairs: effects on physiological activity and adaptive value to a desert shrub. Oecologia 37:183–200.

Engler, A., and K. Prantl. 1887–1915. *Die Naturlichen Pflanzenfamilien*. 23 vols. Leipzig.

Evans, Raymond A., and James A. Young. 1977. Bitterbrush germination with constant and alternating temperatures. J. Range Mgmt. 30:30–32.

——— and ———. 1977. Weed control–revegetation systems for big sagebrush–downy brome rangelands. J. Range Mgmt. 30:331–336.

Everett, Richard L., Richard O. Meewig, Paul T. Tueller, and Raymond A. Evans. 1977. Water potential in sagebrush and shadscale communities. Northwest Science 51:271–281.

Fautin, R. W. 1946. Biotic communities of the northern desert shrub biome in western Utah. Ecol. Monographs 16:251–310.

Fleming, C. E., M. R. Miller, and L. R. Vawter. 1922. The spring rabbitbrush (*Tetradymia glabrata*), a range plant poisonous to sheep. Nev. Agr. Exp. Sta. Bull. 101.

Flowers, Seville. 1934. Vegetation of the Great Salt Lake region. Bot. Gaz. 95:353–418.

Fosberg, M. A., and M. Hironaka. 1964. Soil properties affecting the distribution of big and low sagebrush communities in southern Idaho. Amer. Soc. Agron. Spec. Pub. 5:230–236.

Freeman, D. C., L. G. Klikoff, and K. T. Harper. 1976. Differential resource utilization by the sexes of dioecious plants. Science 191:597–599.

Fulton, Robert E., and F. Lynn Carpenter. 1979. Pollination, reproduction, and fire in California *Arctostaphylos*. Oecologia 38:147–157.

Gasto, Juan M. 1969. Comparative Autecological Studies of *Eurotia lanata* and *Atriplex confertifolia*. Thesis (Ph.D.), Utah State Univ., Logan.

Gates, D. H., L. A. Stoddart, and C. W. Cook. 1956. Soil as a factor influencing plant distribution on salt deserts of Utah. Ecol. Monographs 26:155–175.

Goodwin, Duwayne L. 1956. Autecological Studies of *Artemisia tridentata* Nutt. Thesis (Ph.D.), State College of Washington, Pullman.

Gordon, John C., and Jeffrey O. Dawson. 1979. Potential uses of nitrogen-fixing trees and shrubs in commercial forestry. Bot. Gaz. 140 (suppl.): S88–S90.

Grieve, M. 1971. *A Modern Herbal*. 2 vols. Dover Publications, New York.

Hanson, Craig A. 1962. Perennial *Atriplex* of Utah and the Northern Deserts. Thesis (M.S.), Brigham Young Univ., Provo.

Holbo, H. Richard, and Hugh N. Mozingo. 1965. The chromatographic characterization of *Artemisia*, section *tridentatae*. Amer. J. Bot. 52:970–978.

Howell, John T. 1971. A new name for "winterfat." Wasmann J. Biol. 29:105.

Jackson, Gemma. 1934. The morphology of the flowers of *Rosa* and certain closely related genera. Amer. J. Bot. 21:453–466.

Jones, R. 1970. *The Biology of* Atriplex: *Studies of the Australian Arid Zone.* Commonwealth Scientific and Industrial Research Organization, Canberra, Australia.

Kartesz, John T., and Rosemarie Kartesz. 1980. *A Synonymized Checklist of the Vascular Flora of the United States, Canada, and Greenland.* Vol. 2: *The Biota of North America.* Univ. of North Carolina Press, Chapel Hill.

Kay, Burgess L., James A. Young, Catherine M. Ross, and Walter L. Graves. 1977. Rubber rabbitbrush. Mohave Revegetation Notes 3, Agronomy and Range Science, Univ. of California, Davis.

———, C. M. Ross, and W. L. Graves. 1977. Hop-sage. Ibid. 6.

———, ———, and ———. 1977. White burrobush. Ibid. 7.

———, ———, and ———. 1977. Bush peppergrass. Ibid. 10.

———, ———, and ———. 1977. Cooper's desert thorn. Ibid. 13.

———, Charles R. Brown, and Walter L. Graves. 1977. Fourwing saltbush. Ibid. 17.

———, C. M. Ross, W. L. Graves, and C. R. Brown. 1977. Gray ephedra and green ephedra. Ibid. 19.

———, ———, and ———. 1977. Winterfat. Ibid. 20.

———, J. A. Young, C. M. Ross, and W. L. Graves. 1977. Desert peach. Ibid. 21.

Keeley, Jon E. 1977. Seed production, seed populations in soil, and seedling production after fire for two congeneric pairs of sprouting and non-sprouting chaparral shrubs. Ecology 58:820–829.

——— and P. H. Zedler. 1978. Reproduction of chaparral shrubs after fire: a comparison of sprouting and seeding strategies. Amer. Midl. Nat. 99: 142–161.

Klemmedson, J. O. 1979. Ecological importance of actinomycete-nodulated plants in the western United States. Bot. Gaz. 140 (suppl.): S91–S96.

Lanner, Ronald M. 1983. *Trees of the Great Basin: A Natural History.* Univ. of Nevada Press, Reno.

Love, Lesley D., and N. E. West. 1972. Plant moisture stress patterns in *Eurotia lanata* and *Atriplex confertifolia.* Northwest Science 46:44–51.

McArthur, E. Durant. 1977. Environmentally induced changes of sex expression in *Atriplex canescens*. Heredity 38:97–103.

McConnell, B. R., and G. A. Garrison. 1966. Seasonal variation of available carbohydrates in bitterbrush. J. Wildlife Mgmt. 30:168–172.

McHugh, Tom. 1972. *The Time of the Buffalo*. Univ. of Nebraska Press, Lincoln.

McMinn, Howard E. 1970. *An Illustrated Manual of California Shrubs*. Univ. of Calif. Press, Berkeley and Los Angeles.

McPhee, John. 1981. *The Pine Barrens*. Farrar, Straus, Giroux, New York.

Meinzer, C. E. 1927. Plants as indicators of ground water. USGS Water Supply Paper 577.

Millspaugh, Charles F. 1892. *American Medicinal Plants*. Yorston and Company, Philadelphia.

Mooney, H. A., and M. West. 1964. Photosynthetic acclimation of plants of diverse origin. Amer. J. Bot. 51:825–827.

Moore, Russell T. 1971. Transpiration of *Atriplex confertifolia* and *Eurotia lanata* in Relation to Soil, Plant, and Atmospheric Moisture Stresses. Thesis (Ph.D.), Utah State Univ., Logan.

Moss, E. H. 1940. Interxylary cork in *Artemisia* with a reference to its taxonomic significance. Amer. J. Bot. 27:762–768.

Munz, Philip A., and David D. Keck. 1959. *A California Flora*. Univ. of Calif. Press, Berkeley and Los Angeles.

Nord, Eamor C. 1965. Autecology of bitterbrush in California. Ecol. Monographs 35:307–334.

Passey, H. B., and V. K. Hugie. 1962. Sagebrush on relict ranges in the Snake River plains and northern Great Basin. J. Range Mgmt. 15:273–278.

Peattie, Donald C. 1953. *A Natural History of Western Trees*. Houghton Mifflin, Boston.

Peterson, K. M., and W. W. Payne. 1973. The genus *Hymenoclea* (Compositae: Ambrosieae). Brittonia 25:243–256.

Pope, C. Lorenzo. 1976. A Phylogenetic Study of the Suffrutescent Shrubs in the Genus *Atriplex*. Thesis (Ph.D.), Brigham Young Univ., Provo.

Quick, Clarence R. 1935. Notes on the germination of *Ceanothus* seeds. Madrono. 3:135–140.

———. 1959. *Ceanothus* seeds and seedlings on burns. Madrono. 15:79–81.

——— and A. S. Quick. 1961. Germination of *Ceanothus* seeds. Madrono. 16:23–30.

Reifschneider, Olga. 1964. *Biographies of Nevada Botanists, 1844–1963*. Univ. of Nevada Press, Reno.

Reveal, James L. 1979. Biogeography of the intermountain region. Mentzelia 4.

Robertson, Joseph H. 1947. Responses of range grasses to different intensities of competition with sagebrush (*Artemisia tridentata*). Ecology 28:1–16.

——— and P. B. Kennedy. 1954. Half-century changes on northern Nevada ranges. J. Range Mgmt. 7:117–121.

——— and Howard P. Cords. 1957. Survival of rabbitbrush, *Chrysothamnus* sp., following chemical, burning, and mechanical treatments. J. Range Mgmt. 10:83–89.

———. 1983. Greasewood (*Sarcobatus vermiculatus* [Hook.] Torr.). Phytologia 54:309–324.

Roundy, B. A., J. A. Young, and R. A. Evans. 1981. Phenology of salt rabbitbrush (*Chrysothamnus nauseosus* ssp. *consimilis*) and greasewood (*Sarcobatus vermiculatus*). Weed Sci. 29:448–454.

Sampson, Arthur W., and Beryl S. Jesperson. 1963. California range brushlands and browse plants. Calif. Agr. Exp. Sta. Manual 33.

Schlatterer, E. F., and E. W. Tisdale. 1969. Effects of litter of *Artemisia*, *Chrysothamnus*, and *Tortula* on germination and growth of 3 perennial grasses. Ecology 50:869–873.

Shantz, H. L. 1925. Plant communities in Utah and Nevada. U.S. Natl. Museum Contr., U.S. Natl. Herbarium 25:15–23.

Shaver, Gaius R. 1978. Leaf angle and light absorptance of *Arctostaphylos* species (Ericaceae) along environmental gradients. Madrono. 25:133–138.

Shreve, F. 1942. The desert vegetation of North America. Bot. Rev. 8:195–264.

Smith, J. G. 1900. Fodder and forage plants, exclusive of grasses. U.S. Dept. of Agr., Div. Agrost. Bull. 2, rev.

Stebbins, G. Ledyard. 1972. Evolution and diversity of aridland shrubs. In *Wildland Shrubs: Their Biology and Utilization*, Forest Service, U.S. Dept. of Agr., General Technical Report INT-1, August: 111–120.

Steingraeber, D. A. 1982. Heterophylly and neoformation of leaves in sugar maple (*Acer saccharum*). Amer. J. Bot. 69:1277–1282.

———. 1984. Heterophylly in *Acer glabrum* Torr. Abstracts, Annual Meeting Bot. Soc. of Amer., August.

Strickler, G. S. 1956. Factors Affecting the Growth of Whitesage (*Eurotia lanata*). Thesis (M.S.), Univ. of Nevada, Reno.

Strother, John L. 1974. Taxonomy of *Tetradymia* (Compositae: Senecioneae). Brittonia 26:177–202.

Stutz, Howard C., and L. K. Thomas. 1964. Hybridization and introgression in *Cowania* and *Purshia*. Evolution 18:183–195.

———, J. M. Melby, and G. K. Livingston. 1975. Evolutionary studies of *Atriplex*: a relict gigas diploid population of *Atriplex canescens*. Amer. J. Bot. 62:236–245.

———. 1979. Explosive evolution of perennial *Atriplex* in western North America. In *Intermountain Biogeography: A Symposium*, Great Basin Naturalist Memoirs 2:161–168.

———, C. L. Pope, and S. C. Sanderson. 1979. Evolutionary studies of *Atriplex*: adaptive products from the natural hybrid, 6N *A. tridentata* × 4N *A. canescens*. Amer. J. Bot. 66:1181–1193.

Sweeney, J. R. 1956. Responses of vegetation to fire: a study of herbaceous vegetation following chaparral fires. Univ. Calif. Publ. Bot. 28:143–250.

Tabler, R. D. 1964. The root system of *Artemisia tridentata* at 9500 feet in Wyoming. Ecology 45:633–636.

Tarrant, R. F. 1961. Stand development and soil fertility in a Douglas fir–red alder plantation. Forest Sci. 7:238–246.

——— and J. M. Trappe. 1971. The role of *Alnus* in improving the forest environment. Plant Soil, spec. vol.: 335–348.

Tidestrom, Ivar. 1925. *Flora of Utah and Nevada*. U.S. Natl. Museum Contr., U.S. Natl. Herbarium 25.

Ting, Irwin P. 1961. An Ecological Study of *Dalea polyadenia* Torr. Thesis (M.S.), Univ. of Nevada, Reno.

Train, Percy, James R. Henrichs, and W. Andrew Archer. 1941. Medicinal uses of plants by Indian tribes of Nevada. Contrib. toward a flora of Nevada 45, Agr. Research Service, U.S. Dept. of Agr., Beltsville, Md.

Tueller, Paul T., and Raymond A. Evans. 1969. Control of green rabbitbrush and big sagebrush with 2,4-D and picloram. Weed Sci. 17:233–235.

Turner, W. J. No date. *A Treasury of English Wildlife*. Chanticleer Press, New York.

Vasek, Frank C., H. B. Johnson, and D. H. Eslinger. 1975. Effects of pipeline construction on creosote bush scrub vegetation of the Mohave Desert. Madrono. 23:1–64.

———. 1980. Creosote bush: long-lived clones in the Mojave Desert. Amer. J. Bot. 67:246–255.

Von Marilaun, A. K. 1895. *The Natural History of Plants*. Trans. F. W. Oliver. 2 vols. Blackie and Son, London.

Ward, G. H. 1953. *Artemisia*, section *seriphidium*, in North America. Contrib. Dudley Herbarium 4:155–205.

Watson, Sereno. 1871. *Botany*. U.S. Geological Exploration of the Fortieth Parallel. Vol. 5. Government Printing Office, Washington, D.C.

Wells, P. V. 1961. Succession in desert vegetation on streets of a Nevada ghost town. Science 134:670–671.

———. 1969. The relation between mode of reproduction and extent of speciation in woody genera of the California chaparral. Evolution 23:264–267.

Welsh, Stanley L., and Glen Moore. 1973. *Utah Plants*. 3d ed. Brigham Young Univ. Press, Provo.

West, Marda L. 1969. Physiological Ecology of Three Species of *Artemisia* in the White Mountains of California. Thesis (Ph.D.), Univ. of Calif., Los Angeles.

White, W. N. 1932. A method of investigating ground water supplies based on discharge by plants and evaporation from soil. USGS Water Supply Paper 659A:1–105.

Williams, Stephen E. 1979. Vesicular-arbuscular mycorrhizae associated with actinomycete-nodulated shrubs, *Cercocarpus montanus* Raf. and *Purshia tridentata* (Pursh) DC. Bot. Gaz. 140 (suppl.): S115–S119.

Williamson, R. L. 1967. Productivity of red alder in western Oregon and Washington. In *Biology of Alder*, Proc. Sympos. Northwest Scientific Assoc., 40th Annual Meeting, Pullman, Washington: 287–292.

Windle, Leaford C. 1960. An Investigation of the Ecology of Six Perennial Species of the Salt-desert Shrub Type in Southern Idaho. Thesis (M.S.), Univ. of Idaho, Moscow.

Windward, Alma H. 1970. Taxonomic and Ecological Relationships of the Big Sagebrush Complex in Idaho. Thesis (Ph.D.), Univ. of Idaho, Moscow.

Wood, Benjamin W. 1966. An Ecological Life History of Budsage in Western Utah. Thesis (M.S.), Brigham Young Univ., Provo.

Wood, M. K., R. W. Knight, and J. A. Young. 1976. Spiny hop-sage germination. J. Range Mgmt. 29:53–56.

Young, James A., and Raymond A. Evans. 1974. Population dynamics of green rabbitbrush in disturbed big sagebrush communities. J. Range Mgmt. 27:127–132.

—— and ——. 1974. Phenology of *Chrysothamnus viscidiflorus* subspecies *viscidiflorus*. Weed Sci. 22:469–475.

—— and ——. 1976. Stratification of bitterbrush seeds. J. Range Mgmt. 29:421–425.

——, ——, and B. L. Kay. 1977. Ephedra seed germination. Agron. J. 69:209–211.

—— and ——. 1978. Germination requirements as determinants of species composition of *Artemisia* rangeland communities. Proc. First Intl. Rangeland Congress: 366–369.

INDEX

Abscisic acid, 278
Aceraceae, 217
Acer, 218
   glabrum, 218, 220
      var. *diffusum,* 221
      var. *douglasii,* 218
      var. *torreyi,* 221
   tripartitum, 218
Achene, 94, 157, 285
Actinomorphic, 238
Actinomycete, 38, 211
Aemaeodera, 299
Alder, 10, 36, 37
Alder buckthorn, 214
Alderleaf mountain-mahoghany, 152
Alderleaf sarvisberry, 148
Alkali rabbitbrush, 292
Alkali seepweed, 90
Allelopathy, 238, 276
Allenrolfea, 43
   occidentalis, 43, 44
Alnus, 37
   incana, 41
   tenuifolia, 36, 37
Altered andesite buckwheat, 103, 104

Ambrosia, 304
Amelanchier, 145
   alnifolia, 145, 146
   pallida, 145
   pumila, 145
   utahensis, 145
American dogwood, 199, 200
American Medicinal Plants, 16, 27
Anacardiaceae, 223
Anderson, C. L., 171, 232
Anderson's desert thorn, 232
Antelope bitterbrush, 175
Apache plume, 159
Arborvitae, 17
Archer, W. A., 193, 247
Arctostaphylos, 123
   nevadensis, 126
   patula, 123, 125
Artemisia, 256
   arbuscula, 256, 258
      var. *nova,* 256, 309
   cana, 263
      ssp. *bolanderi,* 263
   spinescens, 265, 266
   tridentata, 270, 274

Artemisia tridentata (continued)
    ssp. *tridentata*, 282
    ssp. *vaseyana*, 281
    ssp. *wyomingensis*, 281
Asteraceae, 255
Aster family, 255
Athel tamarisk, 110
Atriplex, 46
  bonnevillensis, 66
  canescens, 46, 47
    ssp. *aptera*, 46
  confertifolia, 52, 53
  falcata, 65
  lentiformis, 60
  torreyi, 60, 61
  tridentata, 63, 64

Bailey, L. H., 86, 106, 154
Bailey's greasewood, 3, 51, 80, 84, 85, 229
Bakker, E., 287
Barberry, 27
Barberry family, 26
Bark striping, 126
Barneby, R. C., 191
Baum, B. R., 106
Bearberry honeysuckle, 246
Beatley, J., 178
Beech family, 31
Beetle, A. A., 263
Benson, L., 110, 206
Berberidaceae, 26
Berberine, 28
Berberis, 27
  repens, 27, 29
Bessey, C., 116
Betulaceae, 35
Big greasewood, 80
Bighead disease, 309
Big sagebrush, 3, 256, 257, 270, 274
Big sagebrush community, 5, 75, 256
Billings, D., 1, 57, 67, 175, 281

Birch family, 35
Bissell, H., 180
Bitterbrush, 3, 11, 175, 176
Bitter cherry, 172, 173, 214
Bitterroot, 182
Bittersweet, 206
Bittersweet family, 203
Blackbrush, 3, 157, 171
Black cottonwood, 40
Black currant, 134
Black elderberry, 249
Black greasewood, 80, 229
Black sage, 182, 309
Black sagebrush, 256, 257, 309
Black twinberry, 246
Bloomer's goldenbush, 294
Blueberry, 129
Blue elderberry, 249
Bonneville saltsage, 66
Brassicaceae, 118
Brickell, J., 285
Brickellbush, 283
Brickellia, 283
  californica, 286
  grandiflora, 286
  microphylla, 283, 284
  oblongifolia var. *linifolia*, 285
Brigham tea, 19
Broom snakeweed, 301
Buckbrush, 182, 208
Buckthorn, 214
Buckthorn family, 207
Buckwheat family, 91
Buckwheat shrub, 51
Budsage, 265
Bud sagebrush, 3, 265, 266
Buffaloberry, 195
Burning bush, 206
Burrobrush, 157, 304
Bur sage, 314
Bush chinquapin, 3, 32, 33
Bush peppergrass, 119, 120

Bush pickleweed, 43
Button brush, 265

Cacti, 7
Caldwell, M. M., 70, 275
California brickellbush, 283, 286
California buckbrush, 208
California buckwheat, 92
Canadian buffaloberry, 197
Caprifoliaceae, 242
Carpenter, F. L., 123
Cascara sagrada, 214
Castanea pumila, 32
Castanopsis, 32
   chrysophylla, 34
   sempervirens, 32, 33
Catalog of Nevada Flora, 171
Catclaw horsebrush, 315
Ceanothus, 208
   cordulatus, 209, 212
   cuneatus, 208
   integerrimus, 209
   velutinus, 208, 210
      var. *laevigatus*, 213
      var. *lorenzii*, 213
   X *lorenzii*, 213
Celastraceae, 203
Cell potential, 276
Ceratoides, 67
   lanata, 67
Cercis, 247
Cercocarpus, 149
   betuloides, 152
   intricatus, 149, 150
   ledifolius, 149
   montanus, 152
Chamaebatia, 156
Chamaebatiaria, 154
   millefolium, 154, 155
Chamiza, 51
Chaparral, 11
Chenopodiaceae, 42

Chenopodium album, 54
Chico, 232
Chokecherry, 10, 174
Christensen, E., 106
Chromosomes, 48, 55, 261
Chrysothamnus, 287
   albidus, 292, 299
   nauseosus, 287, 288, 290
      ssp. *consimilis*, 82
      ssp. *hololeucus*, 83
   parryi, 292, 293
      ssp. *monocephalus*, 294
   viscidiflorus, 292, 296, 298
      var. *tortifolia*, 297
Chuang, F., 231
Clawson, A. B., 309
Cliffrose, 159, 160, 181
Cocklebur, 304
Coevolution, 56
Coffeeberry, 174, 214
Coleogyne, 157
   ramosissima, 157
Common rabbitbrush, 289
Coniferous forest zone, 3
Cook, C. W., 67
Cooper's desert thorn, 231
Cords, H., 289
Cornaceae, 198
Cornelian cherry, 202
Cornus, 199
   florida, 201
   mas, 202
   nuttallii, 201
   sericea, 199, 200
      ssp. *occidentalis*, 199
   stolonifera, 202
Corokia, 201
Cotton horsebrush, 3, 317
Coville, F. V., 51, 131
Cowan, J., 161
Cowania, 159
   mexicana var. *dubia*, 161

Cowania mexicana (continued)
  var. *stansburiana*, 159, 160, 181
Coyote willow, 113, 115
Crabapple bush family, 206
Crassulacean acid metabolism, 7
Creambush, 162
Creeping barberry, 27, 29
Creeping snowberry, 253
Creosote bush, 169, 287, 307
Creosote bush community, 75
Cronquist, A., 3, 58, 201
Crossomataceae, 206
Cruciferae, 121
Cupressaceae, 13
Curlleaf mountain-mahoghany, 149
Currant, 3, 27, 134
Currant family, 133
Cuticle, 6
Cyclopedia of American Horticulture, 106, 154
Cypress, 17
Cypress family, 13

Dalea polyadenia, 187
Daleae Imagines, 191
Darrow, R., 110, 206
Darwin, C., 55
Davis, C., 126
De Bell, D. S., 40
Deerbrush, 182, 209
DePuit, E. J., 275
De Quille, D., 80
Desert bitterbrush, 181
Desert blite, 3, 87, 88
Desert molly, 77
Desert peach, 168, 170, 232
Desert sage, 238
Desert snowberry, 253
Desert sweet, 154
Desert thorn, 229
Dichogamy, 83

Die Naturlichen Pflanzenfamilien, 114
Diettert, R. A., 276
Dina, S., 278
Dioecious, 20, 48, 49, 50, 55
Diploid, 49, 50, 261
Dogwood, 199
Dogwood family, 198
Doten, S. B., 289, 299
Double honeysuckle, 245, 246
Drosera, 283
Durrant, S., 108
Dwarf juniper, 14, 15
Dwarf maple, 218, 220
Dwarf ninebark, 166
Dwarf sagebrush, 256, 258

Earle, A. M., 183
Eckert, R. E., 68
Ecotype, 50, 57
Elaeagnaceae, 194
Elaeagnus angustifolia, 197
Elderberry, 247, 248
Engler, A., 114
Engler-Prantl system, 114
Ephedra, 8, 19
  antisyphilitica, 23
  nevadensis, 19
    var. *aspera*, 25
    var. *nevadensis*, 25
  viridis, 19, 21, 22, 24
    var. *viridis*, 23
    var. *viscida*, 23
Ephedraceae, 18
Ericaceae, 122
Eriogonum, 92
  fasciculatum, 92
  heermannii, 98, 99
  kearneyi, 92, 93
  lobbii var. *robustum*, 103, 104
  microthecum, 96, 97
  sphaerocephalum, 100, 101
  wrightii ssp. *subscaposum*, 100, 102

Eslinger, D. H., 306
Euonymus, 206
European speckled alder, 41
Eurotia, 72
   ceratoides, 67
Evans, R. A., 23, 82, 175, 178, 297, 299

Fabaceae, 186
Fagaceae, 31
Fallugia paradoxa, 159
Fautin, R. W., 82
Fern bush, 154, 155
Fire, 11, 123, 179, 208
Fire-bush, 79
Flora Americae Septentrionalis, 182
Flora of Utah and Nevada, 149, 158
Flowering Plants and Ferns of Arizona, 95
Flowering dogwood, 201
Flowers, S., 43, 106, 108
Fly honeysuckle, 246
Forselles, J. H., 204
Forsellesia, 204
   nevadensis, 204, 205
Fosberg, M. A., 259
Four-winged saltbush, 3, 46, 47
Freeman, D. C., 20
Frémont, J. C., 1, 90, 121
Fremont dalea, 193
Fulton, R. E., 123

Garrison, G. A., 179
Gasto, J. M., 56, 68
Gates, D. H., 67
Gaylussacia, 132
Geneva, 16
Gerard's Herball, 183, 247
Gigas plants, 49
Glasswort, 43
Glossopetalon nevadense, 204
   spinescens, 206
      var. *aridum*, 206

Goldenbush, 294
Golden currant, 134
Goodwin, D., 279
Gooseberry, 3, 10, 134, 141
Gooseberry family, 133
Goosefoot family, 42
Granite-gilia, 234
Graves, W. L., 169, 289
Gray, A., 73, 171, 204, 231, 295
Gray horsebrush, 308
Grayia, 73
   brandegei, 75
   spinosa, 73, 74
Gray molly, 77
Greasebush, 204
Greasewood, 3, 8, 67, 80, 81, 182
Greasewood association, 5
Greasewood-shadscale association, 58
Great Basin buckwheat, 96, 97
Green ephedra, 3, 19, 21, 22, 24
Greenleaf manzanita, 3, 123, 125
Green molly, 3, 77, 78
Green rabbitbrush, 3, 292, 296, 298
Grey, Z., 238
Grieve, M., 247
Grossularia, 139
Grossulariaceae, 133, 139
Gutierrezia, 301
   sarothrae, 301, 302
Gymnosperm, 19

Halogeton, 68
Halophyte, 9, 109
Handbook of Flower Pollination, 140
Hanson, C. A., 51, 65, 66
Haplopappus, 294
   bloomeri, 294
   nanus, 294
   suffruticosus, 294
Harper, K. T., 20
Hatch-Slack pathway, 59
Hatton, J. H., 127

Heath family, 122
Heermann's buckwheat, 98, 99
Henrichs, J. R., 193, 247
Herbaceous plants, 5
Hexaploid, 65, 262
Hironaka, H., 259
Historia Naturalis, 113
Hoary sagebrush, 263
Holbo, D., 261
Hollygrape, 27
Holmgren, A., 3, 58
Holmgren, N., 3, 58
Holodiscus, 162
   discolor, 162
   dumosus, 162, 163
   microphyllus, 164
Honeysuckle, 243
Honeysuckle family, 242
Hopsage, 3, 51, 73, 74
Hormay, A. L., 178
Horsebrush, 308
Howell, J. T., 72
Huckleberry, 132
Huffman, W. T., 309
Hugie, V. K., 259, 260
Hymenoclea, 304
   monogyra, 307
   salsola, 304, 305
      var. *patula*, 307
      var. *pentalepis*, 307
      var. *salsola*, 307

Ice Age, 9
Illustrated Manual of California Shrubs, An, 134
Incense cedar, 17
Inferior ovary, 147, 249
Inkberry, 246
Intermountain Flora, 3, 58
Introgression, 181–182
Involucre, 96, 100
Iodine bush, 8, 43, 44

Irregular flowers, 238, 240
Island Called California, An, 287

Jack-in-the-pulpit, 20
Jenever, 16
Jesperson, B. S., 75
Johnson, H. B., 306
Jointfir, 19
Jointfir family, 18
Jones, M., 193
Jujube family, 213
Juneberry, 148
Juniperus, 15
   communis, 14, 15
      var. *depressa*, 16
      var. *montana*, 16

Kartesz, J., 167, 231, 251
Kartesz, R., 167
Kay, B. L., 23, 76, 169, 289
Kearney, T. H., 95
Kearney's buckwheat, 92, 93
Keeley, J. E., 124, 127, 211
Kennedy, P. B., 294, 299
Kern greasewood, 43
King's dalea, 193
Klikoff, L. G., 20
Knight, R. W., 75
Knobcone pine, 11
Knobloch, I., 181
Knuth, P., 140
Koch, W. D. J., 79
Kochia, 77
   americana, 77, 78
   prostrata, 79
   scoparia, 79

Lahontan, Lake, 9
Lamb's-quarters, 54
Lamiaceae, 237
Lanner, R. M., 149
Lemonade sumac, 224

Lenscale, 60
Lepidium, 119
   fremontii, 119, 120
   montanum, 119
     var. *canescens*, 119
Leptodactylon, 234
   pungens, 234, 235
Letters, 270
Lewis, M., 136, 264
Lewisia, 182
Linnaea borealis, 254
Linnaeus, C., 108, 254
Little greasewood, 58, 84
Little greasewood–shadscale association, 58
Littleleaf brickellbush, 283, 284
Littleleaf horsebrush, 3, 8, 70, 308, 310 311, 313
Littleleaf mountain-mahoghany, 149, 150
Livingston, G. K., 49
Lodgepole pine, 11
Lodgepole pine–mountain hemlock zone, 3
Longspine horsebrush, 316
Lonicera, 243
   conjugialis, 245, 246
   involucrata, 243, 244
   utahensis, 243
Lonitzer, A., 246
Low sagebrush, 256
Lycium, 229
   andersonii, 232
   barbarum, 232
   cooperi, 231
   halimifolium, 232
   shockleyi, 84, 229, 230

McArthur, E. D., 48
McConnell, B. R., 179
McHugh, T., 195
McMinn, H., 134

McPhee, J., 129
Mahonia, 30
Ma-huang, 19
Manna, 110
Manzanita, 123, 211
Maple family, 217
Marijuana, 20
Marilaun, K., 165
Matchweed, 301
Matrimony vine, 232
Medlar, 148
Meinzer, O., 82
Melby, J. M., 49
Mespilus germanica, 148
Millspaugh, Charles F., 16, 27
Mint family, 237
Modern Herbal, A, 247
Mooney, H. A., 278
Mojave antelope brush, 181
Monoecious, 20, 48, 55
Moore, R., 70
Moquin-Tandon, C., 73
Mormon tea, 19
Mountain alder, 5, 36, 37, 39
Mountain dogwood, 201
Mountain mahoghany, 149
Mountain peppergrass, 119
Mountain snowberry, 251
Mountain whitethorn, 209, 212
Muller, H., 140
Mustard family, 118
Mycorrhizae, 180

Natural History of Plants, 165
Natural History of Western Trees, A, 37
Navajo ephedra, 23
Nebraska currant, 197
Neoteny, 306
Nevada, A Guide to the Silver State, 168
Nevada ephedra, 19
Nevada through Rose-Colored Glasses, 193
Nightshade family, 232

Ninebark, 166
Nitrogen fixation, 38, 180, 211
Nord, E. C., 177
Nuttall, T., 99
Nuttall's buckwheat, 96
Nuttall's horsebrush, 310, 316
Nuttall's saltbush, 68

Oak family, 31
Oblongleaf brickellbush, 285
Ocean spray, 162, 163
O'Donnell, B., 226
Oleaster family, 194
Oregon boxwood, 206
Oregon grape, 10

Paleozoic era, 4
Pappus, 291
Parish, S., 254
Parish's snowberry, 251
Parry, C. C., 295
Parry's rabbitbrush, 293
Passey, H. B., 259, 260
Payne, W., 304
Pea family, 186
Pearlbush, 304
Peattie, D. C., 37
Peebles, R., 95
Peterson, K., 304
Philodendron, 16
Phlox, 234
Phlox family, 233
Photosynthesis, 6, 7, 59, 231, 257, 275
Phreatophyte, 83, 271
Physocarpus, 166
  alternans, 166
Phytochrome, 280
Pickleweed, 43
Pigeonberry, 148
Pine Barrens, The, 129
Pinemat manzanita, 126
Pinyon–juniper zone, 3, 15, 19, 75

Plateau gooseberry, 3, 141, *142*
Pliny, 113
Poison ivy, 226
Poison oak, 226
Polecat bush, 224
Polemoniaceae, 233
Polygonaceae, 91
Polyploid, 261
Pope, C. L., 63
Poplar, 10, 117
Populus, 117
Prantl, K., 114
Prickly phlox, 234, *235*
Prunus, 168
  andersonii, 168, *170*, 232
  emarginata, 172, *173*
Psorothamnus, 187
  fremontii, 193
  kingii, 193
  polydenius, 187, *188*
    var. *jonesii*, 193
  spinosus, 191
Purple sage, 238, *239*
Pursh, F. T., 182, 264
Purshia, 175
  glandulosa, 181
  tridentata, 175, *176*

Quail brush, 60
Quick, A., 208
Quick, C., 208
Quininebrush, 182

Rabbit berry, 197
Rabbitbrush, 52, 67, 287
Rabbit thorn, 232
Rabbitsage, 300
Radwan, M. A., 40
Ragweed, 304
Range Plant Handbook, 286
Red alder, 40
Redbud, 247

Red elderberry, 248, 249
Red osier, 199
Red sage, 77
Regular flowers, 238, 251
Reveal, J., 3, 58
Rhamnaceae, 207
Rhamnus, 214
  cathartica, 214
  frangula, 214
  purshiana, 214
  rubra, 214, 215
Rhubarb, 136
Rhus, 224
  diversiloba, 226
  glabra, 227
  trilobata, 224, 225
  verniciflua, 227
Ribes, 134
  aureum, 134, 135
    var. *gracillimum*, 136
  cereum, 137, 138
  nigrum, 134
  sativum, 134
  velutinum, 141, 142
    var. *glanduliferum*, 141
Riboidoxylon, 143
Riders of the Purple Sage, 238
Robertson, J. H., 83, 271, 272, 289
Robinson, T. W., 84
Rock buckwheat, 100, 101
Rock-spiraea, 162
Rocky Mountain maple, 218
Rolfe, A., 45
Root planing, 289
Root sprout, 11, 124, 178
Rosa, 183
  woodsii, 183, 184
Rosaceae, 144
Rose family, 144
Ross, C. M., 169, 289
Roundy, B., 82
Rubber rabbitbrush, 3, 287, 288, 290

Rubber weed, 294
Russian olive, 197
Rydberg, P. A., 161

Sage, 238
Sagebrush, 11, 52, 67, 256, 263, 265, 270
Sagebrush-grass zone, 3, 9
Sagebrush zone, 3, 19
Salicaceae, 112
Salicin, 117
Salicornia, 43
Salix, 113
  exigua, 113, 115
Saltbush, 46, 67
Saltgrass, 8
Salt rabbitbrush, 83
Saltsage, 51, 63, 64
Salverform flowers, 253
Salvia, 238
  dorii, 238, 239
    var. *argentea*, 241
    var. *carnosa*, 241
    var. *dorrii*, 241
Samara, 221
Sambucus, 247, 248
  cerulea, 247
  racemosa, 248, 249
    var. *melanocarpa*, 249
Sampson, A. W., 75
Sanderson, S. C., 63
Sarcobatus, 80
  baileyi, 58, 80, 84, 85
  vermiculatus, 80, 81
Saskatoon, 148
Schlatterer, E. F., 276
Sedum, 261
Sequoia, 10
Serviceberry, 3, 10, 145, 146
Service tree, 148
Settler's tea, 19
Shadblow, 148

Shadbush, 148
Shadscale, 1, 3, 51, 52, 53, 67, 70, 82
Shadscale community, 75
Shadscale desert, 5, 9
Shadscale-greasewood association, 67
Shadscale zone, 1
Shantz, H. L., 67
Shaver, G., 126
Shepherd, J., 197
Shepherdia, 195
   argentea, 195, 196
   canadensis, 197
   rotundifolia, 197
Shockley's desert thorn, 84, 229, 230
Shortspine horsebrush, 315
Siberian elm, 5
Sickle saltsage, 65
Sierra coffeeberry, 3, 174, 214, 215
Sierra mountain misery, 156
Silver buffaloberry, 195, 196
Silverleaf, 197
Silver sagebrush, 263
Singlehead goldenbush, 294
Skunkberry, 246
Skunkbush, 224
Slender buckwheatbrush, 96
Smith, G. E. P., 110
Smith, J. G., 72
Smokebush, 3, 8, 187, 188
Smoke tree, 191
Smooth sumac, 227
Snakeweed, 3, 301, 302
Snowberry, 3, 251, 252
Snowbrush, 208
Sodium, 68, 109
Solanaceae, 228
Sorbaria, 156
Sorbus domestica, 148
Spineless hopsage, 75
Spiny greasebrush, 8, 204, 205
Spiny hopsage, 73, 74
Spiny sagebrush, 265

Spiraea, 156
Spring sagebrush, 265
Squawberry, 224
Squawbush, 224, 225
Squaw tea, 19
Squawthorn, 232
Stebbins, G. L., 58, 181
Steingraeber, D., 219
Stephanomeria, 314
Sticky laurel, 208
Stoddart, L. A., 67
Stolon, 27
Stomate, 7, 8
Strother, J., 309, 315, 317
Stutz, H. C., 49, 63, 181
Suaeda, 87
   fruticosa, 90
   torreyana, 87, 88
      var. *ramosissima*, 90
Sumac family, 223
Sun-Dials and Roses of Yesterday, 183
Superior, Lake, 9
Superior ovary, 249
Sweeney, J. R., 208
Symphoricarpos, 251, 252
   acutus, 253
   longiflorus, 253
   oreophilus, 251
   parishii, 251
Synangy, 83
Synonymized Checklist of the Vasucular Flora of the United States, Canada and Greenland, A, 167

Tabler, R. D., 271
Tamaricaceae, 105
Tamarisk, 8, 106, 107
Tamarisk family, 105
Tamarix, 106, 107
   aphylla, 110
   chinensis, 108
   mannifera, 110

parviflora, 108
ramosissima, 108
Tarrant, R. F., 40
Tasselflower, 286
Terpenes, 238, 272
Tetradymia, 308
   axillaris, 316
   canescens, 308
   comosa, 317
   glabrata, 310, 311, 313
   nuttallii, 316
   spinosa, 315
   tetrameres, 317
Tetraploid, 49
Thomas, K., 181
Thornber, J. J., 110
Three-lobed sumac, 224
Tidestrom, I., 108, 149, 158
Time of the Buffalo, The, 195
Ting, I., 187
Tisdale, E. W., 276
Tobacco brush, 8, 208, 210
Tomato family, 228
Torrey, J., 62, 89, 161
Torrey saltbush, 60, 61
Toxicodendron diversilobum, 226
   radicans, 226
Train, A., 193
Train, P., 193, 247, 300
Transpiration, 6, 7
Trappe, J. M., 40
Trees and Shrubs of the Southwestern Deserts, The, 110, 206
Trees of the Great Basin, 149
Tueller, P., 299
Turpentine-weed, 301
Twain, M., 270
Twinberry, 243, 244

Utah honeysuckle, 243
Utah mortonia, 206

Vaccinium, 129
   uliginosum ssp. *occidentale*, 129, 130
Varnish leaf ceanothus, 208
Vasek, F., 169, 306

Wafer sagebrush, 51
Waisel, Y., 110
Ward, G., 261
Washoe Rambles, 80
Watson, S., 12, 67
Wax currant, 134, 137, 138
West, M. L., 256, 278
Western blueberry, 129, 130
Western golden currant, 134, 135
Western mountain-mahoghany, 152
Western serviceberry, 145, 146
Wheat rust, 27
White burrobush, 304, 305
White, E. C., 131
White, W., 82
White-flowered rabbitbrush, 292, 299
White-pine blister rust, 27, 141
Whitesage, 67
Wild buckwheat, 92
Wild oleaster, 197
Wild rose, 183, 184
Williams, S., 180
Willow, 10, 113
Willow family, 112
Windle, L., 71
Windward, A., 281
Winterfat, 51, 67, 69
Winterfat association, 67
Wire-lettuce, 314
Wolfberry, 232
Wood, B. W., 267
Wood, M. K., 75
Woodbury, A., 108
Wright's buckwheat, 100, 102

Xanthium, 304
Xanthocephalum, 303

Xerophyte, 273

Yellowbrush, 300
Yellow pine–white fir zone, 3
Yellowsage, 300
Yellow top, 301

Young, J. A., 23, 50, 71, 75, 82, 137, 169, 175, 178, 297

Zedler, P. H., 124, 211
Zygomorphic, 238

Italic page numbers indicate illustrations.

SHRUBS OF THE GREAT BASIN WAS DESIGNED AND ILLUSTRATED BY CHRISTINE RASMUSS STETTER, EDITED BY HOLLY CARVER, AND PROOFREAD BY MARY HILL. COMPOSITION IS IN GOUDY OLD STYLE BY G&S TYPESETTERS, AUSTIN, TEXAS. THE BOOK WAS PRINTED ON MOHAWK SUPERFINE AND WARREN'S LUSTRO OFFSET ENAMEL BY PARAGON PRESS, SALT LAKE CITY, AND BOUND BY HILLIER BINDERY. PRODUCTION SUPERVISION BY RICK STETTER AND RICHARD WORKMAN.